POLYMERS FOR ADVANCED TECHNOLOGIES

Processing Characterization and Applications

POLYMERS FOR ADVANCED TECHNOLOGIES

Processing Characterization and Applications

Edited by
Gennady E. Zaikov DSc, Liliya I. Bazylyak, DSc
and Jimsher N. Aneli, DSc

Apple Academic Press

TORONTO   NEW JERSEY

© 2013 by
Apple Academic Press Inc.
3333 Mistwell Crescent
Oakville, ON L6L 0A2
Canada

Apple Academic Press Inc.
1613 Beaver Dam Road, Suite # 104
Point Pleasant, NJ 08742
USA

First issued in paperback 2021

Exclusive worldwide distribution by CRC Press, a Taylor & Francis Group

ISBN 13: 978-1-77463-266-6 (pbk)
ISBN 13: 978-1-92689-534-5 (hbk)

Library of Congress Control Number: 2012951952

Library and Archives Canada Cataloguing in Publication

Polymers for advanced technologies: processing characterization and applications/edited by Gennady E. Zaikov, Liliya I. Bazylyak, and Jimsher N. Aneli.

Includes bibliographical references and index.
ISBN 978-1-926895-34-5

1. Nanocomposites (Materials). 2. Polymers. I. Zaikov, G. E.
(Gennadi ̄ i Efremovich), 1935- II. Bazylyak, Liliya I III. Aneli, J. N

| TA418.9.C6P64 2013 | 620.1'18 | C2012-906365-7 |

About the Editors

Gennady E. Zaikov, DSc

Gennady E. Zaikov, DSc, is Head of the Polymer Division at the N. M. Emanuel Institute of Biochemical Physics, Russian Academy of Sciences, Moscow, Russia, and a professor at Moscow State Academy of Fine Chemical Technology, Russia, as well as a professor at Kazan National Research Technological University, Kazan, Russia. He is also a prolific author, researcher, and lecturer. He has received several awards for his work, including the the Russian Federation Scholarship for Outstanding Scientists. He has been a member of many professional organizations and on the editorial boards of many international science journals.

Liliya I. Bazylak, DSc

Liliya I. Bazylak, DSc, is Professor and Senior Staff Scientist at the Physical Chemistry of Combustible Minerals Department at the Institute of Physical Organic Chemistry and Coal Chemistry of the National Academy of Sciences of Ukraine, in Lviv, Ukraine. She is a specialist in the field of chemical kinetics, chemistry and physics of polymers and polymer composites.

Jimsher N. Aneli, DSc

Jimsher N. Aneli, DSc, is Professor at the Institute of Machine Mechanics at Georgian Technical University of Tbilisi State University in Tbilisi, Georgia. He is a scientist in the field of chemistry and the physics of oligomers, polymers, composites, and nanocomposites.

Contents

List of Contributors

Ol'ga V. Afonicheva
A. N. Nesmeyanov Institute of Organoelement Compounds Russian Academy of Sciences, 28 Vavilov str. Moscow-119991, Russia.

L. F. Akhmetshina
Izhevsk State Technical University Studencheskaya St. 7, Izhevsk-426069, Russia.
OJSC "Izhevsk Electromechanical Plant "Kupol" Pesochnaya St. 3, Izhevsk-426033, Russia.

V. Z. Aloev
Kabardino-balkarian State Agricultural Academy, Nal'chik–360030, Tarchokov st., 1 a, Russian Federation.

Mikhail I. Buzin
A. N. Nesmeyanov Institute of Organoelement Compounds Russian Academy of Sciences, 28 Vavilov str. Moscow-119991, Russia.

M. A. Chashkin
Izhevsk State Technical University.
Izhevsk Electromechanical Plant.
BRHE Centre of Chemical Physics and Mesoscopy, Udmurt Scientific Centre, UD, RAS

O. I. Demchyna
Department of Physico-chemistry of Combustible Minerals L. M. Lytvynenko Institute of Physico-organic Chemistry and Coal Chemistry NAS of Ukraine Naukova St., 3a, Lviv-79059.

B. Zh. Dzhangurazov
CRR "TD PolyChemGroups", Barklay st., 18/19, Moscow-121309, Russian Federation.

A. K. Haghi
University of Guilan, Rasht, Iran.

B. A. Izmailov
A. N. Nesmeyanov Institute of Organoelement Compounds Russian Academy of Sciences, ul. Vavilova 28, Moscow-119991 Russia.

N. V. Khokhriakov
Izhevsk State Agricultural Academy, Basic Research and Educational Center of Chemical Physics and Mesoscopy, Udmurt Scientific Center, Ural Division, Russian Academy of Science, Russia, Izhevsk-426000.

V. I. Kodolov
Izhevsk State Technical University Studencheskaya St. 7, Izhevsk-426069, Russia.
OJSC "Izhevsk Electromechanical Plant "Kupol" Pesochnaya St. 3, Izhevsk-426033, Russia.
BRHE Centre of Chemical Physics and Mesoscopy, Udmurt Scientific Centre, UD, RAS, Studencheskaya St. 7, Izhevsk-426069, Russia.

G. A. Korablev
Izhevsk State Agricultural Academy, Basic Research and Educational Center of Chemical Physics and Mesoscopy, Udmurt Scientific Center, Ural Division, Russian Academy of Science, Russia, Izhevsk-426000.

Stefan Kubica
G. V. Plekhanov Russian Economic University 36 Stremyannyi way, Moscow 117997 Russia
Institut Inzynierii Materialow Polimerowych I Barwnikow, 55 M. Sklodowskiej-Curie str., 87-100 Torun, Poland.

Elena I. Lozinskaya
A.N. Nesmeyanov Institute of Organoelement Compounds Russian Academy of Sciences, 28 Vavilov str. Moscow-119991, Russia

T. Z. Lygina
Central Scientific-Research Institute of Geology Non-Ore Minerals, Zinin Street 4, 420097 Kazan, Russia.

G. M. Magomedov
Dagestan State Pedagogical University, Yaragskii st., 57, Makhachkala-367003, Russian Federation.

Yu. G. Medvedevskikh
Department of Physico-chemistry of Combustible Minerals L. M. Lytvynenko Institute of Physico-organic Chemistry and Coal Chemistry NAS of Ukraine Naukova St., 3a, Lviv-79059.

O. V. Mikhailov
Kazan National Research Technological University.

A. K. Mikitaev
Kabardino-Balkarian State University, Chernyshevskiy st., 173, Nal'chik-360004, Russian Federation.

N. I. Naumkina
Central Scientific-Research Institute of Geology Non-Ore Minerals, Zinin Street 4, 420097 Kazan, Russia.

Anatoly N. Neverov
G. V. Plekhanov Russian Economic University 36 Stremyannyi way, Moscow-117997 Russia.

Elena L. Pekhtasheva
G. V. Plekhanov Russian Economic University 36 Stremyannyi way, Moscow-117997 Russia.

E. N. Rodlovskaya
A. N. Nesmeyanov Institute of Organoelement Compounds Russian Academy of Sciences, ul. Vavilova 28, Moscow-119991 Russia.

H. V. Romaniuk
Lviv Polytechnic National University Bandera St., 12, Lviv-79013, Ukraine.

S. N. Salazkin
Institute of Russian Academy of Sciences A. N. Nesmeyanov Institute of Organoelement Compounds Russian Academy of Sciences, Moscow, Russia ul. Vavilova 28, Moscow-119991 Russia.

Alexander S. Shaplov
A. N. Nesmeyanov Institute of Organoelement Compounds Russian Academy of Sciences, 28 Vavilov str. Moscow-119991, Russia

V. V. Shaposhnikova
Institute of Russian Academy of Sciences A.N. Nesmeyanov Institute of Organoelement Compounds Russian Academy of Sciences, Moscow, Russia ul. Vavilova 28, Moscow-119991 Russia.

Ol'ga V. Sinitsyna
A. N. Nesmeyanov Institute of Organoelement Compounds Russian Academy of Sciences, 28 Vavilov str. Moscow-119991, Russia

V. V. Trineeva
Izhevsk State Technical University.
Izhevsk Electromechanical Plant.
BRHE Centre of Chemical Physics and Mesoscopy, Udmurt Scientific Centre, UD, RAS.
Institute of Applied Mechanics, Ural Division, Russian Academy of Science.

I. I. Tsiupko
N. M. Emanuel Institute of Biochemical Physics Russian Academy of Sciences Kosygin St., 4, Moscow-119334, Russia.

M. A. Vakhrushina
Izhevsk Electromechanical Plant.

Yu. M. Vasilchenko
Izhevsk State Technical University.
Izhevsk Electromechanical Plant.
BRHE Centre of Chemical Physics and Mesoscopy, Udmurt Scientific Centre, UD, RAS

Yu. G. Vasiliev
Izhevsk State Agricultural Academy, Basic Research and Educational Center of Chemical Physics and Mesoscopy, Udmurt Scientific Center, Ural Division, Russian Academy of Science, Russia, Izhevsk-426000.

V. A. Vasnev
A.N. Nesmeyanov Institute of Organoelement Compounds Russian Academy of Sciences, ul. Vavilova 28, Moscow-119991 Russia.

Tat'yana V. Volkova
A.N. Nesmeyanov Institute of Organoelement Compounds Russian Academy of Sciences, 28 Vavilov str. Moscow-119991, Russia.

Yakov. S. Vygodskii
A. N. Nesmeyanov Institute of Organoelement Compounds Russian Academy of Sciences, 28 Vavilov str. Moscow-119991, Russia.

Kh. Sh. Yakh'yaeva
Dagestan State Pedagogical University, Yaragskii st., 57, Makhachkala 367003, Russian Federation.

Yu. G. Yanovskii
Institute of Applied Mechanics of Russian Academy of Sciences, Leninskiy pr., 32 a, Moscow-119991, Russian Federation.

I. Yu. Yevchuk
Department of Physico-chemistry of Combustible Minerals L. M. Lytvynenko Institute of Physico-organic Chemistry and Coal Chemistry NAS of Ukraine Naukova St., 3a, Lviv-79059.

Olesya N. Zabegaeva
A. N. Nesmeyanov Institute of Organoelement Compounds Russian Academy of Sciences, 28 Vavilov str. Moscow-119991, Russia.

G. E. Zaikov
N. M. Emanuel Institute of Biochemical Physics of Russian Academy of Sciences, Kosygin st., 4, Moscow-119334, Russian Federation.

A. I. Zakharov
Izhevsk Electromechanical Plant.

Z. M. Zhirikova
Kabardino-Balkarian State Agricultural Academy, Nal'chik–360030, Tarchokov st., 1 a, Russian Federation.

List of Abbreviations

APCL	Anionic ring-opening polymerization of ε-caprolactam
CNF	Carbon nanofibers
CNT	Carbon nanotubes
Co/C NC	Cobalt/carbon nanocomposite
CHEC	Cold-hardened epoxy composition
Cu/C NC	Copper/carbon nanocomposite
DMF	N,N-dimethylformamid
DSC	Differential scanning calorimetry
DMSO	Dimethylsulfoxide
CL	ε-caprolactam
ED	Electron diffraction
EMD	Electron microdiffraction
EDR	Epoxy-diane resin
ER	Epoxy resin
EPS	Expanded polystyrene
FS	Fine suspensions
GHG	Greenhouse gas
HDPE/CaCO$_3$	High density polyethylene/calcium carbonate
HSC	High-strength concrete
ILs	Ionic liquids
IROM	Inverse rule of mixture
LLDPE	Linear low density polyethylene
LDPE	Low density polyethylene
LLDPE/MMT	Low density polyethylene/Na$^+$-montmorillonite
MMT	Montmorillonite
MBT	Mechanical-biological treatment
MSW	Municipal solid waste
NS	Nanostructures
Ni/C NC	Nickel/carbon nanocomposite
OPC	Ordinary Portland cement
PCA	Polycaproamide
PC	Polycarbonate
PDPP	Polydiphenylenephthalide
PE	Polyethylene
PEPA	Polyethylene polyamine
PCM	Polymeric composite materials
PP	Polypropylene
PVDF	Polyvinilidene fluoride
RH	Rice husk ash
ROM	Rule of mixture
SF	Silica fume

SEI	Spatial-energy exchange interactions
SEP	Spatial-energy parameter
TU	Technical condition
TGA	Thermo gravimetric analysis
TEM	Transmission electron microscopy
WTE	Waste-to-energy

Preface

This book is primarily designed for students of bachelor, diploma, and master courses of materials science, materials technology, plastic technology, mechanical engineering, process engineering, and chemical engineering. It can be used by students, teachers at universities, and colleges for supplementary studies in the disciplines of chemistry and industrial engineering. The polymer testing methods are essential to the development and application of biomedical or nanostructured materials.

This book is an outgrowth of an organized compilation of the notes the authors have used to teach advanced courses of polymers for many years.

The authors have long held the view that the lack of knowledge of the fundamental aspects of polymer materials is a serious shortcoming in undergraduate as well as graduate science and engineering education. This is especially important in our present society because the use of polymeric materials pervades our experience both in our daily lives and in our engineering profession. Still, the basic thrust of undergraduate and graduate education to some degree is in the areas of mechanical and civil engineering is toward traditional materials of metal, concrete, etc. Until about twenty-five years ago, undergraduate textbooks on materials had little coverage of polymers. Today, many materials texts have several chapters on polymers, but in general, the thrust of such courses is toward metals. Even the polymer coverage that exists now stresses the analysis of polymers using the same procedures as for metals and other materials and, therefore, often misleads the young engineer on the proper design of engineering plastics.

Thus, it is not surprising that some structural products made from polymers are often poorly designed and do not have the durability and reliability of structures designed with metallic materials.

With the publication of this book, we hope to not only serve the important task of training young scientists in physical and materials oriented disciplines but also to make a contribution to further the education of professional polymer testers, design engineers, and technologists.

— **Gennady E. Zaikov, DSc**

1 Updates on Application of Silver Nanoparticles

*N. I. Naumkina, O. V. Mikhailov,
and T. Z. Lygina*

CONTENTS

1.1 INTRODUCTION

Small sized colloid elemental silver particles formed in gelatin layers during the development of silver halide photographic emulsions find a mention about more than 40 years ago [1]. In a number of studies, there are indications of the existence of a separate phase of the elemental silver consisting of nanoparticles and are received as a result of photochemical reduction of Ag(I) salts. It appeared lately in the literature [2-11]. It has been noted that [10, 11], during the development of gelatin layers of silver halide photographic emulsions by alkaline water solutions containing tin(II) dichloride and some inorganic or organic substance forming stable coordination compounds with Ag(I), the formation of elemental silver occurs too. However, the gelatin layer is either tinged brown or red but not black color due to the fact that it takes place at standard development by using hydroquinone developers. It is significant to note that with the increase of optical densities of the gelatin layer with elemental silver, red tone coloring of gelatin layer becomes more and more clearly expressed. The similar phenomenon takes place when instead of silver halide AgHal in gelatin matrix, there is such silver(I) compound as silver(I) hexacyanoferrate(II) $Ag_4[Fe(CN)_6]$. Whether this totality of particles is a novel phase of elemental silver? Or is it only a variety of known phases of the given simple substance? These questions remain unanswered till now and deserve special consideration.

1.2 EXPERIMENTAL

As the initial material to obtain silver-containing gelatin-immobilized matrix implants (GIM), X-ray film Structurix D-10 (Agfa-Gevaert, Belges) was chosen. Samples of the given film (which actually was nothing but AgHal-GIM) having format 20×30 cm^2 were exposed to X-ray radiation with an irradiation dose at range 0.05-0.50 Röntgen. These exposed samples were further subjected to processing according to the following technology [10, 11]:

- Development in D-19 standard developer as it was indicated in [10, 11], for 6 min at 20-25°C;
- Washing with running water for 2 min at 20-25°C;
- Fixing in 25% water solution of sodium trioxosulphidosulphate(VI) (Na$_2$S$_2$O$_3$) for 10 min at 20-25°C;
- Washing with running water for 15 min at 18-25°C.

First three stages of standard processing (development, washing, and fixing) were carried out at non-actinic green-yellow light, and final washing at natural light. The samples of GIM containing elemental silver (Ag-GIM) were processed according to the next following technology:

(1) Oxidation in water solution containing (gl^{-1})

Potassium hexacyanoferrate(III)	50.0
Potassium hexacyanoferrate(II)	20.0
Potassium hydroxide	10.0
Sodium trioxocarbonate(IV) (Na$_2$CO$_3$)	5.0
Water	up to 1000 ml

for 6 min at 20-25°C;

(2) Washing with running water for 2 min at 20-25°C;
(3) Reduction in water solution containing (gl^{-1})

Tin(II) chloride	50.0
Sodium N,Nэ-ethylene diaminetetra acetate	35.0
Potassium hydroxide	50.0
Reagent formed water soluble complex with Ag(I)	1.0-100.0
Water	up to 1000 ml

for 1 min at 20-25°C;

(4) Washing with running water for 15 min at 18-25°C;
(5) Drying for 2-3 h at 20-25°C.

As complex forming reagents that form water soluble complexes with Ag(I), ammonia NH$_3$, potassium thiocyanate KSCN, sodium trioxosulphidosulphate(VI) Na$_2$S$_2$O$_3$, ethanediamine-1,2 H$_2$N–CH$_2$–CH$_2$–NH$_2$, 2-aminoethanol H$_2$N–CH$_2$–CH$_2$–OH, and 3-(2-hydroxyethyl)-3-azapenthanediol-1,5 N(CH$_2$–CH$_2$–OH)$_3$ were used. At the first stage of given processing of Ag-GIM obtained, conversion of Ag-GIM into Ag$_4$[Fe(CN)$_6$]-GIM occurred; in the second stage, reduction of Ag$_4$[Fe(CN)$_6$]-GIM with Sn(II) from elemental silver took place. And so, peculiar "re-precipitation" of elemental silver into gelatin matrix occurred incidentally.

An isolation substance from Ag-GIM was carried out by means of influence on them of water solutions of some proteolytic enzymes (for example, trypsin or *Bacillus mesentericus*) destroying the polymeric carrier of GIM (gelatin) and the subsequent separation of a solid phase from mother solution according to a technique described in

[16]. The substances isolated thus from GIM were further analyzed by X-ray diffraction method by using spectrometer D8 Advance (Bruker, Germany). A scanning was carried out in interval from 3 to 65° × 2θ, a step was 0.05 × 2θ.

Calculation of intensities of reflexes (I) and inter-plane distances (d) was carried out with application of standard software package EVA. Theoretical XRD spectra (X-ray patterns) were calculated under Powder Cell program described in [17, 18]. Optical density Ag-GIM was measured by means of Macbeth TD504 photometer (Kodak, USA) in range 0.1-5.0 units with accuracy of $\pm$ 2% (rel.).

1.3 RESULTS

Already at visual observation over a course of transformation process of $Ag_4[Fe(CN)_6]$-GIM to Ag-GIM, the following circumstance attracts its attention. The Ag-GIM received as a result of standard processing of exposed AgHal-GIM, at small optical density (D^{Ag}), has gray color, and at big D^{Ag}, black color. The colored Ag-GIM containing the "re-precipitated" elemental silver varies from black-brown to red, depending on the nature and quantity of complex forming reagent present in the solution.

It is significant, however, that absorption spectra of both initial and the "re-precipitated" elemental silver in visible area do not contain any accurately expressed maxima. Besides, as per rule, optical density Ag-GIM with the "re-precipitated" silver (D^{Ag}) at the same volume concentration of elemental silver (C_{Ag}^{V}) in GIM is essentially more than D^{Ag} values and also it depends on the nature and quantity of complex forming reagent in solution contacting with GIM. Examples of $D = f(D^{Ag})$ и $D = f(C_{Ag}^{V})$ dependence for inorganic and organic reagents are presented in Figures 1-6.

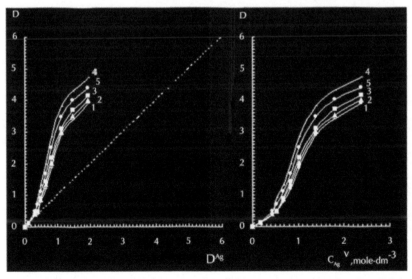

FIGURE 1   Dependence of $D = f(D^{Ag})$ and $D = f(C_{Ag}^{V})$ in reduction process of $Ag_4[Fe(CN)_6] \rightarrow Ag$ using NH_3 at concentration 1.5 gl^{-1} (curve 1), 3.0 gl^{-1} (2), 4.5 gl^{-1} (3), 6.0 gl^{-1} (4), and 7.5 gl^{-1} (5). Optical densities D^{Ag} and D were measured with blue light-filter with a transmission maximum at 450 nm.

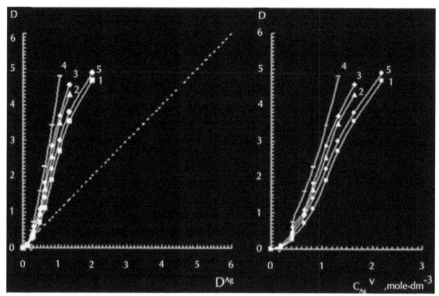

FIGURE 2   Dependence of $D = f(D^{Ag})$ and $D = f(C_{Ag}^{V})$ in reduction process of $Ag_4[Fe(CN)_6] \rightarrow Ag$ using $Na_2S_2O_3$ at concentration 2.0 gl^{-1} (curve 1), 4.0 gl^{-1} (2), 8.0 gl^{-1} (3), 24.0 gl^{-1} (4), and 40.0 gl^{-1} (5). Optical densities D^{Ag} and D were measured with blue light-filter with a transmission maximum at 450 nm.

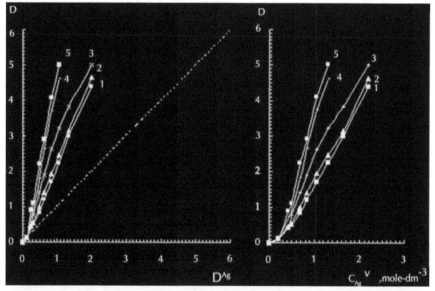

FIGURE 3   Dependence of $D = f(D^{Ag})$ and $D = f(C_{Ag}^{V})$ in reduction process $Ag_4[Fe(CN)_6] \rightarrow Ag$ using KSCN at concentration 2.0 gl^{-1} (curve 1), 4.0 gl^{-1} (2), 8.0 gl^{-1} (3), 24.0 gl^{-1} (4), and 60.0 gl^{-1} (5). Optical densities D^{Ag} and D were measured with blue light-filter with a transmission maximum at 450 nm.

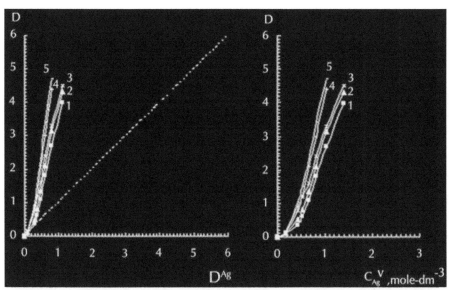

FIGURE 4   Dependence of D = f(D^{Ag}) and D = f(C$_{Ag}$^{V}) in reduction process of Ag$_4$[Fe(CN)$_6$]→Ag using 2-aminoethanol H$_2$N–(CH$_2$)$_2$–OH at concentration 7.5 gl^{-1} (curve 1), 15.0 gl^{-1} (2), 55.0 gl^{-1} (3), 110.0 gl^{-1} (4), and 150.0 gl^{-1} (5). Optical densities D^{Ag} and D were measured with blue light-filter with a transmission maximum at 450 nm.

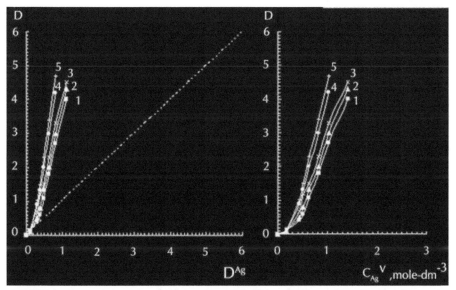

FIGURE 5   Dependence of D = f(D^{Ag}) and D = f(C$_{Ag}$^{V}) in reduction process Ag$_4$[Fe(CN)$_6$]→Ag using ethanediamine-1,2 H$_2$N–(CH$_2$)$_2$–NH$_2$ at concentration 5.0 gl^{-1} (curve 1), 10.0 gl^{-1} (2), 20.0 gl^{-1} (3), 40.0 gl^{-1} (4), and 80.0 gl^{-1} (5). The optical densities D^{Ag} and D were measured with blue light-filter with a transmission maximum at 450 nm.

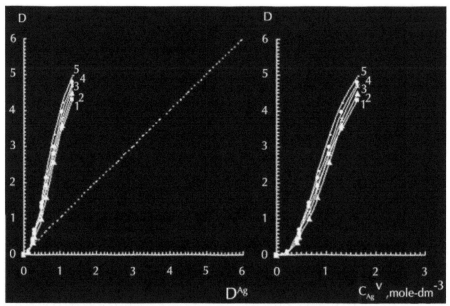

FIGURE 6   Dependence of $D = f(D^{Ag})$ and $D = f(C_{Ag}{}^{V})$ in reduction process of $Ag_4[Fe(CN)_6] \rightarrow Ag$ using 3-(2-hydroxyethyl)-3-azapenthanediol-1, 5 $N(CH_2-CH_2-NH_2)_3$ at concentration 10.0 gl^{-1} (curve 1), 20.0 gl^{-1} (2), 35.0 gl^{-1} (3), 50.0 gl^{-1} (4), and 100.0 gl^{-1} (5). The optical densities D^{Ag} and D were measured with blue light-filter with a transmission maximum at 450 nm.

The D/D^{Ag} value, as a rule, greater than 1.0, and in some cases, it reaches very high values (as in the case of potassium thiocyanate--nearly 5.0). Attention is drawn to the fact that the stronger the color of the gelatin layer with the "re-precipitated" elemental silver, different from the gray-black tones of the gelatinous layer and initially Ag-gelatin-immobilized matrix, the greater is the D/D^{Ag} value. The maximal degree of amplification $(D/D^{Ag})_{max}$ is also very much dependent on the nature of the complex forming reagents (Table 1). The most profound effect on this parameter has etandiamin-1,2 $[(D/D^{Ag})_{max} = 5.80]$, the least severe ammonia, the degree of possibility is also quite high $[(D/D^{Ag})_{max} = 3.40]$. For ammonia, the growth (D/D^{Ag}) value is typical with increasing concentration of NH_3 in reducing solution to a relatively smaller value (~0.30 $moll^{-1}$), after which the optical density D begins to fall (Figure 1). It is noteworthy that red-brown color of gelatin layer attained at the indicated concentration, with further increase in the concentration of ammonia, does not change. Analogous situation occurs in the case of the other two we studied inorganic complex forming reagent--trioxosulfidosulfate(VI) and the thiocyanate anion (Figures 2-3), with the only difference being that in the case of $S_2O_3{}^{2-}$, maximum degree of amplification is achieved with less high in comparison with NH_3 concentration (0.15 $moll^{-1}$), in the case of SCN⁻--at a higher concentration (~0.70 $moll^{-1}$). In this regard, it was quite natural to try the difference marked with different stability of 1:2 complexes formed by Ag(I) with NH_3, $S_2O_3{}^{2-}$ and SCN⁻ (pK = 7.25, 13.32, and 8.39, respectively). However, in the presence of such correlation, the value of concentration indicated for SCN⁻ should be lower than for NH_3, which in fact, is not observed. In reality, for the three organic complex forming reagents studied, molar concentrations at which the maximum value of D/D^{Ag} is reached are significantly greater than those for inorganic complex forming reagents (Table 1). Stability of the complexes of silver(I) with each of these ligands is lower than with NH_3, and the correlation function between these concentrations and the stability of coordination compounds of Ag(I) with given ligands in varying degrees, is still visible. But any explosion term relationship, the stability of the (D/D^{Ag}) values do not see: How it may be easily noticed while comparing the data of Table 1, the maximum degree of amplification decreases in the direction of ethanediamine-1,2 > 2-aminoethanol > SCN⁻ > $S_2O_3^{2-}$ > NH_3 > 3-(2-hydroxyethyl)-3-azapenthane-diol-1,5, while the resistance formed by these ligands complexes with silver(I)--in the direction $S_2O_3^{2-}$ > SCN⁻ > ethanediamine-1,2 > NH_3 > 2-aminoethanol > 3-(2-hydroxyethyl)-3-azapenthanediol-1,5. Thus, the complex is though important but not the sole determinant of the degree of influence of complex forming agents on the redox process considered.

At the first stage of the given process, reaction described by general Equation (1), takes place (in the braces {....}, formulas of gelatin-immobilized chemical compounds have been indicated):

$$4\{Ag\} + 4[Fe(CN)_6]^{3-} \rightarrow \{Ag_4[Fe(CN)_6]\} + 3[Fe(CN)_6]^{4-} \qquad (1)$$

TABLE 1 The maximal (D/D^{Ag}) and pK_s values of Ag(I) complexes for various complex forming reagents.

Complex forming reagent	$(D/D^{Ag})_{max}$	Concentration of complex forming reagent in solution at which reaches $(D/D^{Ag})_{max}$, gl⁻¹ (molel⁻¹)	pK_s of Ag(I) complex having 1:2 composition
NH_3	3.40	4.5 (0.27)	7.25
$Na_2S_2O_3$	4.13	23.7 (0.15)	13.32
KSCN	4.92	69.7 (0.70)	8.39
HO–$(CH_2)_2$–NH_2	5.40	109.8 (1.80)	6.62
H_2N–$(CH_2)_2$–NH_2	5.80	77.0 (1.28)	7.84
$N(CH_2$–CH_2–$OH)_3$	3.93	99.8 (0.67)	3.64

Each of complex forming reagents under examination forms with Ag(I) soluble complex having a metal ion: ligand ratio of 1:2. That is why, formation of silver(I) complex with corresponding CR will occur to some extent when $Ag_4[Fe(CN)_6]$-GIM is at the contact with the solution containing any of complex forming reagent indicated. Gelatin-immobilized silver(I) hexacyanoferrate(II) as well as any of these soluble complexes can participate in the process of reduction with Sn(II). In this connection, proceeding two parallel processes Ag(I) → Ag(0) will take place in contact of $Ag_4[Fe(CN)_6]$-GIM with solution indicated above, containing Sn(II) and complex forming reagent:

• Gelatin-immobilized silver(I) hexacyanoferrate(II) reduction proceeding in a polymer layer,

- Ag(I) complex with complex forming reagent reduction proceeding on interface of phases GIM/solution.

In water solutions at pH = 12-13, Sn(II) is mainly in a form of hydroxo complex $[Sn(OH)_3]^-$. In this connection, general Equation (2) may be offered for the first of these processes:

$$\{Ag_4[Fe(CN)_6]\} + 2[Sn(OH)_3]^- + 6OH^- \rightarrow 4\{Ag\} + 2[Sn(OH)_6]^{2-} + [Fe(CN)_6]^{4-} \quad (2)$$

For the second of these processes, general Equation (3)

$$2[AgL_2]^+ + [Sn(OH)_3]^- + 3OH^- \rightarrow 2Ag + 4L + [Sn(OH)_6]^{2-} \quad (3)$$

In the case of non-charged ligands and general Equation (4)

$$2[AgL_2]^{(2z-1)-} + [Sn(OH)_3]^- + 3OH^- \rightarrow 2Ag + 4L^{z-} + [Sn(OH)_6]^{2-} \quad (4)$$

In the case of negative charged "acid" ligands may be ascribed (L--symbol of ligand and z--its charge). The particles of elemental silver, formed as a result of Equations (3) and (4), theoretically should have smaller sizes than the particles of elemental silver arising in polymer layer of GIM. To be a part of substance immobilized in GIM, these particles should place freely in intermolecular cavities of gelatin layer. Only in this case, they may diffuse in GIM and may be immobilized in gelatin mass.

Gelatin has an extremely high surface area and an extensive system of micropores. The fragment of its structure has been shown in Figure 7; as may be seen, it contains many intermolecular cavities. It may be valued the average size of intermolecular cavity in the gelatin structure [12-15].

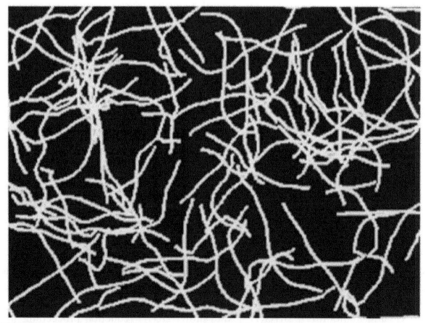

FIGURE 7   The fragment of gelatin structure containing intermolecular cavities.

For example, the volume of polymer layer of GIM (V_{gl}) having area 1 cm^2 and thickness 20 μm is $(1.0 \times 1.0 \times 20 \times 10^{-4})$ $cm^3 = 2.0 \times 10^{-3}$ cm^3, so that the mass of gelatin contained in such a layer, at average value of its density 0.5 gcm^{-3}, is $(0.5 \times 2.0 \times 10^{-3})$ g $= 1.0 \times 10^{-3}$ g. Molecular mass of gelatin (M_{Gel}) is known to be $\sim(2.0-3.0)10^5$ c.u. [12, 13], the number of its molecules in given mass will be $(1.0 \times 10^{-3}/M_{Gel})\cdot(6.02 \times 10^{23}) = (2.0-3.0)10^{15}$. As it was already mentioned above, gelatin molecule in average has length $\sim 28,5000$ pm and diameter $\sim 1,400$ pm, and if it is considered as narrow cylinder, total volume of gelatin molecules V_M will be equal to $(1/4)\pi D^2 h = (1/4)3.14 (285,000 \times 10^{-10}$ cm$)(1,400 \times 10^{-10}$ cm$)^2 = 4.38 \cdot 10^{-19}$ cm^3. In the case of maximal compact arrangement, these molecules occupy total volume equal to 4.38×10^{-19} $cm^3 \times (2.0-3.0)10^{15} = (8.76-13.15)10^{-4}$ cm^3. It may be postulated that the volume of cavities indicated, is equal to total volume of polymer massif minus the volume occupied by gelatin molecules, namely $(2.0 \times 10^{-3} - (8.76-13.15)10^{-4})$ cm^3 that will be in the end $(0.69-1.12)10^{-3}$ cm^3. Then, the average volume of one intermolecular cavity may be found as a quotient from division of their total volume to number of gelatin molecules and, as it may be easily noted, will be $(3.4-5.6)10^{-19}$ $cm^3 = (3.4-5.6)10^{11}$ pm^3. The linear size of such an "average" cavity in the case when it has spherical form, will be equal to $d = (6V/\pi)^{1/3} = [6 \times (3.7-5.6)10^{11}$ $pm^3/3.14]^{1/3} = (89.1-102.2)10^2$ pm; when it has cubic form, equal to $a = V^{1/3} = [(3.7-5.6)10^{11}$ $pm^3]^{1/3} = (71.8-82.4)10^2$ pm. As one can see from these values, these cavities are nanosized. Therefore, only nanoparticles of substance can enter into these cavities. By entering into such cavities, nanoparticles of elemental silver are isolated from each other. In consequence of thereof, their aggregation with each other becomes rather difficult.

With the concentration growth of any of the complex forming reagent mentioned above, the concentration of coordination compounds formed given complex forming reagent with Ag(I) must increase. Correspondingly, the quantity of nanoparticles of the elemental silver formed as a result of reduction of these coordination compounds by $[Sn(OH)_3]^-$ complex should also increase.. In this connection, it may be expected that when concentration of these complex forming reagent in solution increases, the share of nanoparticles contained in the "re-precipitated" elemental silver, should accrue gradually. Thus, at the same concentration nanoparticles of elemental silver owing to their higher dispersion degree in comparison with microparticles should provide higher degree of absorption of visible light (and, accordingly, higher optical density) in polymeric layer GIM. The experimental data presented in Figures 1-6, are in full conformity with the given prediction.

The particles of elemental silver formed as a result of Equations (3) and (4) are one or two nuclear. While it is not enough of them [it occurs, when concentration of Ag(I) complexes on interface of phases, these particles owing to their remoteness from each other have no time to be aggregated. They diffuse in polymeric layer of GIM, and are immobilized without change of their sizes. With the increase of complex forming reagent concentration [and, accordingly, of concentration of Ag(I) complex with given reagent], the quantity of nanoparticles indicates on interface of phases the GIM/ solution accrues. It leads to increase of a number of such particles, diffused into GIM. However, at some rather high concentration complex forming reagent in solution, the effect of aggregation of nanoparticles of elemental silver starts to affect. One and two nuclear particles of elemental silver formed at reduction of corresponding Ag(I) complex, to some extent begin to unite with each other in larger particles. Polynuclear particles of elemental silver resulting such an association are not so mobile and consequently, will not be diffused into polymeric layer of GIM. They will be precipitated in it near to interface GIM/solution (or even to escape as solid phase in the solution contacting with GIM). As a result, rates of an increment of number of one and two nuclear particles of elemental silver with further growth of concentration complex forming reagent begin to be slowed down. Thus, the moment should come inevitably there when the number of similar particles will reach some limiting value. That is why, since certain "threshold" concentration complex forming reagent in solution is there, growth of D^{Ag} values must stop. Moreover, at excess of this "threshold" concentration, certain decrease in D^{Ag} values should begin. The point is that an alignment between number of the aggregated particles and number one and two nuclears with growth of concentration of Ag(I) complex continuously grows and has no restrictions. These polynuclear particles are precipitated in Frontier zone GIM on small depth and form, as a matter of fact, the microparticles of elemental silver formed as a result of reduction of gelatin-immobilized $Ag_4[Fe(CN)_6]$ according to Equation (2). Since D^{Ag} values with increase of complex forming reagent concentration at first increase, they reach a maximum and then decrease.

It may be assumed that "re-precipitated" gelatin-immobilized silver should contain, as a minimum, two phases of the silver particles, one of which is formed by nanoparticles, and another, by microparticles. In order to corroborate the given conclusion, we carried out the analysis of elemental silver isolated from initial Ag-GIM after the end of "re-precipitation" process by X-ray powder diffraction method. X-ray powder diffraction patterns (XRD-patterns) of samples obtained are presented in Figures 8-10. As may be seen from them, XRD-pattern of initial elemental silver with gray-black color of gelatin layer (Figure 8) and XRD-patterns of "re-precipitated" elemental silver (Figures 9-10), rather essentially differ from each other. So, in XRD-patterns of "re-precipitated" elemental silver obtained at an availability of any of studied CR in solution contacting with GIM, there are accurate reflexes having $d = 333.6, 288.5, 166.7$, and 129.1 pm that are absent in XRD-pattern of initial elemental silver. At the same time, reflexes with $d = 235.7, 204.1, 144.4, 123.1$, and 117.9 pm are observed on them. These reflexes are characteristics for the known phase of elemental silver isolated from initial Ag-GIM. In this connection, there are all reasons to believe that the "re-precipitated" elemental silver obtained on using solution containing any of complex forming reagent indicated above, contains at least two structural modifications of elemental silver.

The next curious circumstance attracts its attention: reflexes with $d = 333.6, 288.5, 204.2, 166.7$, and 129.1 pm are rather close to d values of reflexes of silver(I) bromide AgBr (number of card PDF 06-0438, parameter of an elementary cell $a_0 = 577.45$ pm, face-centered lattice, cubic syngonia, $Fm3m$ group of symmetry according to the international classification [14, 15]). In this connection, it may be assumed that the structure of the novel phase contained in "re-precipitated" elemental silver, at least in outline, resembles structure AgBr and its crystal lattice is similar to a lattice of silver(I) bromide where positions of atoms Br occupy atoms of silver.

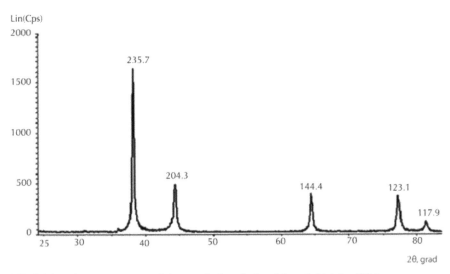

FIGURE 8 The XRD-pattern of elemental silver isolated from initial Ag-GIM.

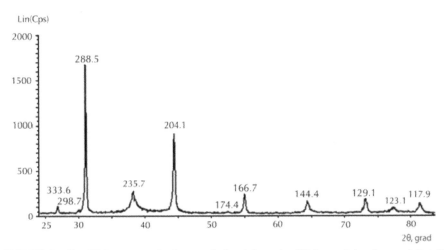

FIGURE 9 The XRD-pattern of substance isolated from Ag-GIM containing "re-precipitated" elemental silver and obtained with using of solution containing Na2S2O3 in concentration 20.0 gl–1.

To answer the question whether the reflexes indicated can belong to elemental silver with such a space structure in principle, theoretical XRD-patterns of assumed structure of elemental silver with the use of program powder cell described in work [18] have been constructed by us. These XRD-patterns are presented in Figure 11. As may be seen from them, the theoretical d values, calculated by us (333.6, 288.7, 204.2, 174.1, 166.7, 144.4, 132.5, 129.1, and 117.9 pm) for specified above structure with an elementary cell parameter $a = 288.72$ pm and *Pm3m* symmetry group, practically coincide with *d* values experimentally observed in XRD-pattern of the "re-precipitated" elemental silver (d = 333.6, 288.5, 166.7, and 129.1 pm).

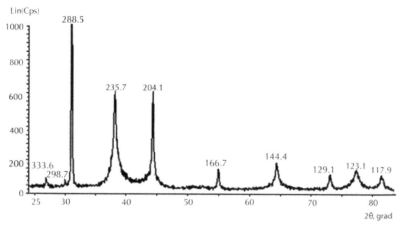

FIGURE 10   The XRD-pattern of substance isolated from Ag-GIM containing "re-precipitated" elemental silver and obtained on using solution containing ethanediamine-1,2 in concentration 20.0 gl–1.

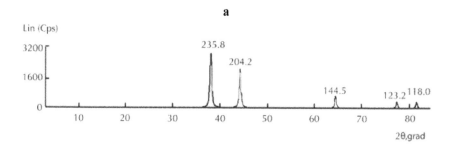

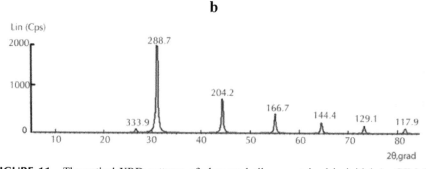

FIGURE 11   Theoretical XRD-patterns of elemental silver contained in initial Ag-GIM (a) and of elemental silver formed in GIM as a result of "re-precipitation" process (b) It should be noted in this connection that d values calculated theoretically for the elemental silver isolated from initial Ag-GIM (235.4, 204.3, 144.5, 123.2, and 118.0 pm), correspond to compact-packed crystal structure having an elementary cell parameter a = 408.62 pm and Fm3m symmetry group, interplane distances in which are 235.7, 204.1, 144.4, 123.1, and 117.9 pm.

Thus, formation of a novel phase of elemental silver, which probably, was not described in the literature up to now, takes place here indeed.

KEYWORDS

- **Complex forming reagents**
- **Gelatin**
- **Gelatin-immobilized matrix**
- **Intermolecular cavity**
- **Light-filter**
- **Nanoparticles**

ACKNOWLEDGMENT

The Russian Foundation of Basic Researches (RFBR) is acknowledged for the financial support of given work (grant No. 09-03-97001).

REFERENCES

1. Skillman, D. G. and Berry, C. R. Effect of particle shape on the spectral absorption of colloid silver in gelatin. *J. Chem. Phys.*, **48**(7), 3297–3304 (1968).
2. Linnert, T., Mulvaney, P., Henglein, A., and Weller, H. Long-lived nonmetallic silver clusters in aqueous solution: Preparation and photolysis. *J. Am. Chem. Soc.*, **112**(12), 4657–4664 (1990).
3. Fedrigo, S., Harbich, W., Butter, J. Collective dipole oscillations in small silver clusters embedded in rare-gas matrices. *Phys. Rev.*, **47**(23), 10706–10715 (1993).
4. Satoh, N., Hasegawa, H., Tsujii, K., and Kimura, K. Photoinduced coagulation of Ag nanocolloides. *J. Phys. Chem.*, **98**(7), 2143–2147 (1994).
5. Sato, T. Ishikawa, T., Ito, T., Yonazawa, Y., Kodono, K., Sakaguchi, T., and Miya, M. *Chem. Phys. Lett.*, **242**(3), 310–314 (1995).
6. Al-Obaidi, A. H. R., Rigbi, S. J. McGarvey, J. J., Wamsley, D. G. Smith, K. W., Hellemans, I., and Snauwaert, J. Microstructural and spectroscopies studies of metal liquidlike films of silver and gold. *J. Phys. Chem.*, **98**(24), 11163–11168 (1994).
7. Ershov, B. G. and Henglein, A. Reduction of Ag$^+$ on polyacrilate chains in aqueous solutions. *J. Phys. Chem., B*, **102**(24), 10663–10666 (1998).
8. Kapoor, S. Surface modification of silver particles. *Langmuir*, **14** (5), 1021–1025 (1998).
9. Sergeev, B. M., Kiryukhin, M. V., and Prusov, A. N. Effect of light on the disperse composition of silver hydrosols stabilized by partially decarboxylated polyacrylate. *Mendeleev Commun.*, **11**(2), 68–69 (2001).
10. Mikhailov, O. V., Guseva, M. V., and Krikunenko, R. I. An amplification of silver photographic images by using of processes changing disperse of image carrier. *Zh. Nauchn. Prikl. Foto-Kinematogr.*, **48**(4), 52–57 (2003).
11. Mikhailov, O. V. Kondakov, A. V., and Krikunenko, R. I. Image intensification in silver halide photographic materials for detection of high-energy radiation by repreciptation of elemental silver. *High Energy Chem.*, **39**(5), 324–329 (2005).
12. Mikhailov, O. V. Reactions of nucleophilic, electrophilic substitution and template synthesis in the metalhexacyanoferrate(II) gelatin-immobilized matrix. *Rev. Inorg. Chem.*, **23**(1), 31–74 (2003).
13. Mikhailov, O. V. Gelatin-Immobilized Metalcomplexes: Synthesis and Employment. *J. Coord. Chem.*, **61**(7), 1333–1384 (2008).

14. Mikhailov, O. V. Self-Assembly of Molecules of Metal Macrocyclic Compounds in Nanore-actors on the Basis of Biopolymer-Immobilized Matrix Systems. *Nanotechnol. Russ.*, **5**(1–2), 18–25 (2010).
15. Mikhailov, O. V. Soft template synthesis of Fe(II,III), Co(II,III), Ni(II) and Cu(II) metalmacro-cyclic compounds into gelatin-immobilized matrix implants. *Rev. Inorg. Chem.*, **30**(4), 199–273 (2010).
16. Mikhailov, O. V. Enzyme-assisted matrix isolation of novel dithiooxamide complexes of nickel(II). *Indian J. Chem.*, **30A**(2), 252–254 (1991).
17. Kraus, W. and Nolze, G. Powder Cell--A program for the representation and manipulation of crystal structures and calculation of the resulting X-ray powder patterns. *J. Appl. Cryst.*, **29**(3), 301–303 (1996).
18. Powder Diffract File. Search Manual Fink Method. Inorganic. USA, Pennsylvania: JCPDS – International Centre for Diffraction Data, 1995 (release 2000).

2 Practical Guide to Nanofiller Aggregation Influence on Nanocomposites

G. V. Kozlov, G. E. Zaikov, and Mikitaev

CONTENTS

2.1 INTRODUCTION

Introduction of different fillers of organic or inorganic origin is one of the most effective and economically profitable methods for enhancement of mechanical properties of polyethylene. Calcium carbonate ($CaCO_3$) is one from such fillers, applied for this purpose for a considerable length of time. It is mainly used at different functions for the production of polyolefines films to increase productivity, giving white dull surface of a film, printing-tracing simplication. However, $CaCO_3$, with particles of the micrometer range, is inert filler and influences weakly on the obtained composites mechanical properties. Application of $CaCO_3$ nanoparticle changes this situation far from being always for the better even at coupling agent using [1-3]. The authors [4] showed that polymer nanocomposites, filled with $CaCO_3$ nanoparticles and relatively low reinforcement degree was due to the interfacial low level adhesion and the indicated aggregation of nanoparticles. Two factors influence on mechanical behavior of particulate-filled polymer nanocomposites, which raises a question on one of the factors predomination or their influence of equal value. The purpose of the present chapter is the problem study on the example of nanocomposites high density polyethylene/calcium carbonate (HDPE/$CaCO_3$).

2.2   EXPERIMENTAL

The HDPE of industrial production with mark 27673 was used as matrix polymer. Nanodimensional $CaCO_3$ in the form of compound with mark Nano-Cal NC-KO117 (China) with particle size of 80 nm and mass contents of 130% was used as nanofiller.

Nanocomposites HDPE/$CaCO_3$ was prepared by components mixing in melt on twin screw extruder, Thermo Haake model, Reomex RTW 25/42, production of German Federal Republic. Mixing was performed at temperature 483493K and screw speed of 1525 rpm for 5 min. Testing samples were obtained by casting under pressure method on casting machine, Test Samples Molding Apparatus RR/TS MP of firm Ray-Ran (Taiwan) at temperature 473K and pressure 8 MPa.

Uniaxial tension mechanical tests have been performed on the samples in the shape of two-sided spade with sizes according to GOST 112 6280. Tests have been conducted on universal testing apparatus Gotech Testing Machine CT-TCS 2000, production of German Federal Republic, at temperature 293K and strain rate $\sim(2\ '\ 10^{-3}\ s^{-1})$.

2.3. RESULTS

In Figure 1, the dependence of elasticity modulus E_n on mass contents of nanofiller W_n for nanocomposites HDPE/$CaCO_3$ is adduced. At $CaCO_3$ contents W_n £ 20 mass % E_n growth at W_n increasing is observed within the limits, typical for polymer composites in general but at $W_n = 30\%$ mass E_n sharp decay occurs, which is unobserved earlier for both microcomposites (polymer composites with filler of micron sizes)[5] and nanocomposites[6]. Such strong effect requires more thorough consideration. Let us study the possible influence of two above indicated factors (nanofiller particles aggregation and interfacial adhesion level) separately and in more detail.

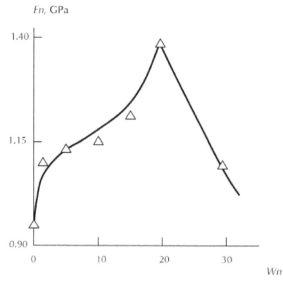

FIGURE 1   The dependence of elasticity modulus E_n on nanofiller mass contents W_n for nanocomposites HDPE/$CaCO_3$.

A nanofiller initial particles aggregation degree can be estimated within the framework of strength dispersive theory [6], where nanocomposite shear yield stress τ_n is determined as follows:

$$\tau_n = \tau_m + \frac{G_n b_B}{\lambda} \qquad (1)$$

where τ_m is polymer matrix shear yield stress, G_n is nanocomposite shear modulus, b_B is Burgers vector, and l is distance between nanofiller particles.

In case of nanofiller particles aggregation, Equation (1) has the following appearance [6]:

$$\tau_n = \tau_m + \frac{G_n b_B}{k(\rho)\lambda}, \qquad (2)$$

where $k(r)$ is aggregation parameter.

The included in the Equations (1) and (2) parameters are determined as follows. The general relationship between normal stress s and shear stress t has the appearance [7]:

$$\tau = \frac{\sigma}{\sqrt{3}} \qquad (3)$$

Young's modulus E and shear modulus G are connected with each other as follows [8]:

$$G = \frac{E}{d_f} \qquad (4)$$

where d_f is nanocomposite structure fractal dimension, which is determined according to the equation [8]:

$$d_f = (d-1)(1+\upsilon), \qquad (5)$$

where d is dimension of Euclidean space, in which fractal is considered (it is obvious that in our case $d = 3$), n is Poisson's ratio, estimated according to the mechanical test results with the aid of equation [9]:

$$\frac{\sigma_Y}{E_n} = \frac{1-2\upsilon}{6(1+\upsilon)}, \qquad (6)$$

where σ_Y is nanocomposite yield stress.

Burgers vector b_B value for polymeric materials is determined according to the relationship [10]:

$$b_B = \left(\frac{60.5}{C_\infty}\right)^{1/2} \text{Å} \qquad (7)$$

where $C_¥$ is characteristic ratio, connected with dimension d_f by the equation [10]:

$$C_\infty = \frac{2d_f}{d(d-1)(d-d_f)} + \frac{4}{3} \qquad (8)$$

The distance between nanofiller non-aggregated particles l can be estimated according to the following equation [6]:

$$\lambda = \left[\left(\frac{4\pi}{3\varphi_n}\right)^{1/3} - 2\right]\frac{D_p}{2}, \qquad (9)$$

where φ_n is nanofiller volume fraction, D_p is its particles diameter ($D_p = 80$ nm). The value φ_n calculation can be fulfilled according to the equation [6]:

$$\varphi_n = \frac{W_n}{\rho_n}, \qquad (10)$$

where ρ_n is nanofiller density, determined from the empirical formula [6]:

$$\rho_n = 0.188\left(D_p\right)^{1/3}, \qquad (11)$$

where D_p is given in nm.

From Equations (2) and (9), aggregation parameter $k(r)$ is very strong, increasing from 5.5 up to 600 within the range of $W_n = 130\%$ mass. Let us consider how such growth $k(r)$ is reflected on nanofiller particles aggregates with diameter D_{ag}. The Equations (911) combination gives the following expression:

$$k(\rho)\lambda = \left[\left(\frac{0.251\pi D_{ag}^{1/3}}{W_n}\right)^{1/3} - 2\right]\frac{D_{ag}}{2} \qquad (12)$$

Allowing, determining at replacement of D_p on D_{ag} real, that is taking into consideration nanofiller particles aggregation, and nanoparticles $CaCO_3$ aggregates diameter D_{ag}. Calculation according to the Equation (12) shows strong increase in D_{ag} (corresponding to $k(r)$ growth) from 320 up to 7,000 nm within the range of $W_n = 130\%$mass. It is important to note that at $W_n = 20$ and 30% mass, the values $D_{ag} = 5$ and 7 nm, that is the considered nanocomposites cease to be as such and turned into microcomposites, even according to the purely formal signs.

Further, the real value ρ_n can be calculated according to the Equation (11) for the aggregated nanofiller and according to the Equation (10)real filling degree ρ_n^{ag}. In Table 1, the comparison of filling degree φ_n^0 for non-aggregated initial $CaCO_3$ and φ_n^{ag}

is adduced. As one can see, the initial nanoparticles $CaCO_3$ aggregation process results into real filling degree φ_n^{ag} essential reduction in comparison with the nominal one φ_n^0. It is obvious that the indicated change φ_n^{ag} explains the behavior of the adduced in Figure 1 of dependence $E_n(W_n)$, since the value φ_n is the main parameter, checking nanocomposite elasticity modulus E_n enhancement in comparison with the corresponding parameter for matrix polymer E_m.

TABLE 1   The comparison of filling degree values φ_n, estimated by different methods, for nanocomposites HDPE/$CaCO_3$

W_n, mass %	φ_n^0	φ_n^{ag}	φ_n^{red}	$\Delta\varphi_n$, %
1	0.012	0.008	0.008	–
5	0.063	0.032	0.035	8.6
10	0.123	0.064	0.068	5.9
15	0.185	0.094	0.100	6.0
20	0.247	0.063	0.071	11.3
30	0.370	0.084	0.097	13.4

For correctness verification of the adduced above filling degree φ_n^{ag} change, which is due to nanoparticles $CaCO_3$ aggregation, the scaling model, applied successfully in the chapter [11], can be used. This model essence consists in the reduction factor α^{n-3} application, connecting real and nominal filling degree (φ_n^{red} and φ_n^0, accordingly):

$$\varphi_n^0 = \alpha^{n-3}\varphi_n^{red},\qquad(13)$$

where a is nanofiller particle (aggregates of particles) size ratio, n is parameter, characterizing nanofiller particle shape, which is accepted equal to 1 for short fibers, 2 for disk-like (flaky) particles, and 3 for spherical particles, φ_n^{red} is a reduced (real) filling degree.

Let us consider the parameters a and n in the Equation (13) choice. The value a should be defined as ratio:

$$\alpha = \frac{D_p}{D_{ag}}\qquad(14)$$

And the parameter n is accepted equal to 2.7 in supposition that $CaCO_3$ nanoparticles' aggregate shape is close to spherical one.

In Table 1, the comparison of filling degree φ_n^{ag} and φ_n^{ag} is adduced, from which their qualitative and quantitative correspondence follow (the average discrepancy of φ_n^{ag} and φ_n^{red} $\Delta\varphi_n$ makes up 7.5%). Therefore, all adduced above results confirmed strong aggregation of $CaCO_3$ nanoparticles at their content growth, the practical point of view is reflected from the study of essential reduction in real filling degree (φ_n^{ag} or φ_n^{red}) of nanocomposites.

Let us consider further the interfacial adhesion level change in nanocomposites HDPE/CaCO$_3$, which estimates the parameter b_a value, allowing clear qualitative gradation of the indicated level. So, the value $b_a = 0$ means interfacial adhesion absence, $b_a = 1.0$ means perfect (by Kerner) interfacial adhesion, and the condition $b_a > 1$ defines nanoadhesion effect [6]. The parameter b_a quantitative estimation can be fulfilled with the aid of the equation [6]:

$$\frac{E_n}{E_m} = 1 + 11\left(1.1\varphi_n^{ag} b_\alpha\right)^{1.7}.$$ (15)

As the estimations according to the Equation (15) have shown, the value b_a for nanocomposites HDPE/CaCO$_3$ changes within the limits of 9.370.84 at W_n growth within the range of 13% mass, which demonstrates interfacial adhesion high level in the indicated nanocomposites (the average value $b_a = 2.93$ corresponds to nanoadhesion effect realization [6]). It is obvious that such a high level of interfacial adhesion is due to the deposition of coupling agent on the basis of stearic acid on the used CaCO$_3$.

Let us consider in conclusion, which of the two described effects (nanofiller initial particles high aggregation degree and interfacial adhesion level) predominates in nanocomposites mechanical behavior. The authors [12] considered three main cases of the dependence of reinforcement degree E_c/E_m (where E_c is composite elasticity modulus) on filler volume contents φ_n. They have shown, that there are dependence $E_c/E_m(\varphi_n)$ of the following main types:

1. Perfect adhesion between filler and polymer matrix, described by Kerner equation, which can be approximated by the following relationship:

$$\frac{E_c}{E_m} = 1 + 11.6\varphi_n - 44.4\varphi_n^2 + 96.3\varphi_n^3;$$ (16)

2. Zero adhesional strength at a large friction coefficient between filler and polymer matrix, which is described by the equation:

$$\frac{E_c}{E_m} = 1 + \varphi_n;$$ (17)

3. Complete absence of interaction and ideal slippage between filler and polymer matrix, when composite elasticity modulus is defined practically by polymer cross section and connected with filling degree by the equation:

$$\frac{E_c}{E_m} = 1 - \varphi_n^{2/3}.$$ (18)

In Figure 2, comparison of the theoretical dependence $E_n/E_m(W_n)$ is adduced where calculation according to the Equations (16) and (17) nominal values φ_n^0, are used for non-aggregated nanofiller, and the corresponding experimental data. This comparison demonstrates clearly prevalent nanoparticles $CaCO_3$ aggregation degree influence on the studied nanocomposites reinforcement degree: at interfacial level higher than perfect one by Kerner, that is at $b_a > 1.0$, the experimental data correspond to the theoretical dependence, calculated according to the Equation (17), that is interfacial adhesion absence or $b_a = 0$. In other words, nanoparticles $CaCO_3$ aggregation, resulting to φ_n^{ag} large reduction with comparison with φ_n^0, levels off interfacial adhesion high average level influence.

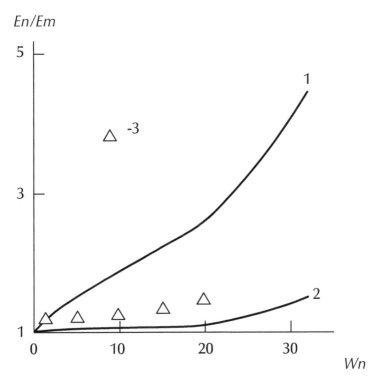

FIGURE 2 Comparison of the theoretical (1, 2) and experimental (3) dependence of reinforcement degree E_n/E_m on nanofiller mass contents W_n for nanocomposites HDPE/$CaCO_3$. The theoretical dependence is calculated as per the Equations (16) (1) and (17) (2) with values φ_n^0 usage.

In the present chapter, the offered structural treatment allows defining nanocomposites' main mechanical characteristics (elasticity modulus E_n and yield stress σ_Y [1-3]) weak enhancement and estimate their limiting magnitudes. So, from the Equations (15) and (16) it follows, that the initial nanoparticles $CaCO_3$ aggregation process suppression, that is the condition $\varphi_n^{ag} = \varphi_n^0 = 0.37$, for nanocomposite HDPE/CaCO$_3$ with $W_n = 30\%$ mass, will give the values $E_n = 3.22$ and 3.90 GPa, accordingly, even at the condition $b_a = 1.0$, that is essentially higher than experimentally obtained value $E_n = 1.09$ GPa.

For nanocomposites HDPE/CaCO$_3$, very weak increase σ_Y at W_n growth is obtained: maximum from 24 up to 29 MPa, that is by 21% only. Such weak enhancement σ_Y explained in Equation (2) analysis: G_n (and E_n) small values and $k(r)$ high magnitudes, that is also due to the initial nanoparticles CaCO$_3$ aggregation. The value $\sigma_Y = 45$ MPa can be reached at aggregation process suppression (even a partial one) and real values $E_n = 2.0$ GPa and $k(r) = 7$.

2.4 CONCLUSION

Therefore, the present chapter results have shown that clearly expressed effect of the initial nanofiller particle aggregation that reduces substantially nanocomposites HDPE/CaCO$_3$ reinforcement degree, leveling interfacial adhesion high level even at using coupling agent. This effect suppression, particularly at high enough nanofiller contents, can improve substantially particulate filled nanocomposites mechanical characteristics. From the practical point of view, aggregation intensification means real filling degree reduction of nanocomposites.

KEYWORDS

- **Euclidean space**
- **High density polyethylene**
- **Microcomposites**
- **Nanoparticles**
- **Polymer nanocomposites**

REFERENCES

1. Osman, A. M., Atallah, A., and Suter, U. W. *Polymer*, **45**(3), 11771183 (2004).
2. Xie, X. L., Liu, Q. X., Li, R. K. Y., Zhou, X. P., Zhang, Q. X., Yu, Z. Z., and Mai, Y. W. *Polymer*, **45**(19), 66656673 (2004).
3. Yang, K., Yang, Q., Li, G., Sun, Y. S., and Feng, D. *Mater. Lett.*, **60**(7), 805809 (2006).
4. Kozlov, G. V., Aphashagova, Z. K., Yanovskii, Y. G., Karnet, Y. N. *Mekhanika, Kompozitsionnykh Materialov i Konstruktsii*, **15**(1), 137148 (2009).
5. Ahmed, S. and Jones, F. R. *J. Mater. Sci.*, **25** (12), 49334942 (1990).
6. Mikitaev, A. K., Kozlov, G. V., and Zaikov, G. E. *Polymer Nanocomposites: Variety of Structural Forms and Applications*. Nova Science Publishers, Inc., New York, p. 319 (2008).
7. Honeycombe, R. W. K. *The Plastic Deformation of Metals*. Edward Arnold Publishers, LTD, London, p. 408 (1968).
8. Bakankin, A. S. *Synergetics of Deformable Body*. Publishers of Ministry Defence SSSR, Moscow, p. 404 (1991).
9. Kozlov, G. V. and Sanditov, D. S. *Anharmonic Effects and Physical-Mechanical Properties of Polymers*. Novosibirsk, Nauka, p. 261 (1994).

10. Kozlov, G. V. and Zaikov, G. E. *Structure of the Polymer Amorphous State.*, Brill Academic Publishers, Utrecht, Boston, p. 465 (2004).
11. Kozlov, G. V., Yanovskii, Y. G., and Karnet, Y. N. *Structure and Properties of Particulate-Filled Polymer Composites: Fractal Analysis.* Al'yanstransatom, Moscow, p. 363 (2008).
12. Tugov, I. I., Shaulov, A. Y., and Vysokomolek. *Soed. B*, **32**(7) 527529 (1990).

3 Practical Hints of Different Aspects of Metal/Carbon Nanocomposites Properties

L. F. Akhmetshina and V. I. Kodolov

CONTENTS

3.1. INTRODUCTION

Nanotechnology is based on self-organization of nanosize particles resulting in the improvement of material properties and obtaining of products and items with unique characteristics. The perspective of the chapter is the modification of silicates and liquid glass with nanocomposites containing metal clusters. The interest to liquid glass is conditioned, first of all, by its ecological friendliness and then followed by production and application simplicity, inflammability and nontoxicity, biological stability, and raw material availability. Nanostructures can be used to improve such characteristics of paints as elasticity, adhesion to the base, hydrophobic behavior, and also helps solving the problems connected with coating flaking off from the base, discoloration, limited color range, and so on. Due to the unique properties of nanostructures, we can apply new properties to silicate paints, for example, to produce the coating, which protects electromagnetic action.

The application of nanostructures in silicate materials to improve their stability to external action is known [1].

The introduction of nanosize structures into the liquid glass allows changing the material behavior qualitatively in a positive way. This possibly occurs in the process of binder supermolecular structure change.

Based on the aforesaid findings, it is important to improve operational properties of compositions on liquid glass basis modifying them with metal/carbon nanocomposites, providing a wider field of their application.

The aim of this chapter is to investigate and analyze the influence of various metal/carbon nanocomposites on changes in the properties of compositions on liquid glass basis.

3.2   EXPERIMENTAL

3.2.1   Nanocomposites Structure

To modify the materials, nickel and iron containing nanocomposites were obtained with low temperature synthesis in nanoreactors of polymeric matrices of poly(vinyl chloride) [2] and poly(vinyl alcohol) [3]. Iron oxide (3) was selected as metal containing phase, which in the process of obtaining nanostructures is reduced to magnetite and nickel oxide, which was also reduced.

To define the sizes, shape, and structure the nanoproduct obtained was investigated with transmission electron microscopy (TEM) and electron diffraction (ED). Iron/carbon nanocomposite (Fe/C NC) is mainly represented as film nanostructures with metal inclusions (Figure 1).

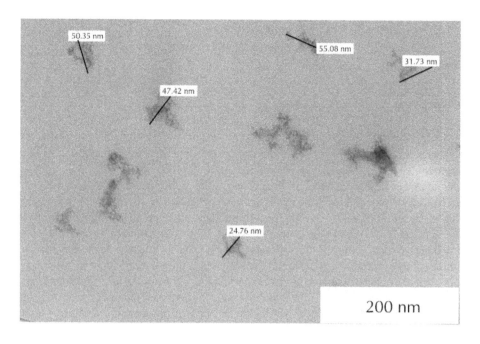

FIGURE 1   *(Continued)*

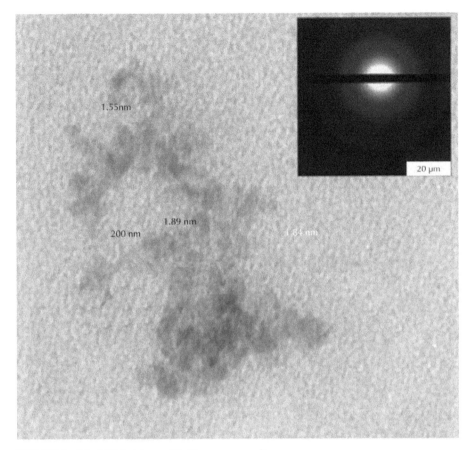

FIGURE 1   The TEM picture of Fe/C nanocomposite.

The particles of metal containing phase have a globular shape and are located between the layers of carbon films connected with them. There are also nanostructures close to a spherical shape. In TEM pictures of Fe/C NC, dark spots characterize metal, light ones carbon films. Electron diffraction patterns are presented as ring reflexes. This indicates the availability of amorphous phase in the sample. The average size of particles constituting the aggregates is 17 nm.

Figure 2 demonstrates large particles and aggregations of nickel clusters. A significant number of particles have the shape close to spherical; however, there are elongated particles as well. Globular film nanostructures with the inclusion of nickel nanoparticles and its compounds are present in the samples. The electron diffraction pattern is presented as ring reflex; however, dot reflexes are also present. This indicates the availability of both crystalline and amorphous phases in the sample. The average size of particles is 11 nm.

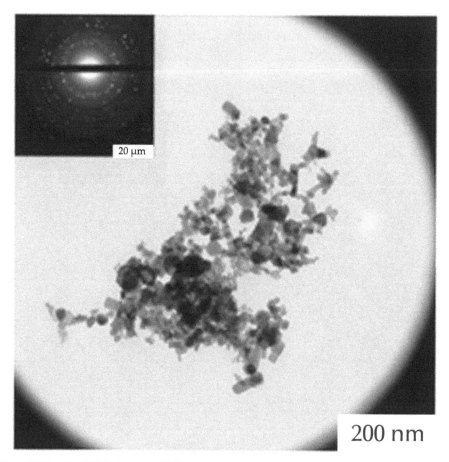

FIGURE 2   The TEM picture of Ni/C nanocomposite.

3.2.2   Equipment and Investigation Technique

1. Spectrophotometry: To define the interaction between liquid glass and nanocomposites, the optical density of liquid glass sample films with nanostructures was investigated. The films were produced applying a thin layer of silicate modified with nanostructures onto transparent film with further drying to remove the moisture. Films without nanocomposites were used as the reference. The investigations were carried out with spectrophotometer KFK-3-01 in the wave range 950350 nm.

2. Ultrasound Processing of Suspensions: Spectrophotometer KFK-3-01 was also used to define the optimal action time interval on nanocomposite fine suspension processed with ultrasound. In the process of selecting the work wavelength (λ) on spectrophotometer KFK-3-01 to define the optimal time interval of ultrasound processing of fine suspension on liquid glass basis, the optimal wavelength of 430 nm was found, as with this wavelength, the optical density of liquid glass with NC before the US processing was 0.818, thus corresponding to the midpoint of the work range of this instrument (work range: optical density (D) = 0.0011.5). At λ = 430 nm, the optical densities of all fine suspensions investigated were found. The optical density was measured for the samples of nanocomposite fine suspension and liquid glass with the concentration of 0.003%, processed in ultrasound bath Sap fir UZV 28 with US power 0.5 kW and frequency 35 kHz. Liquid glass, not modified with nanostructures, was used as a reference sample. The processing time with maximum D was selected as optimal.

3. Investigation of Heat physical Characteristics: Heat capacity of the samples was investigated with dynamic technique measuring the specific heat capacity of solids with c-calorimeter IT-c-400 [4]. The measuring device was based on comparative technique of c-calorimeter with heat meter and adiabatic shell. Temperature delay time on heat meter was measured with stopwatch at room temperature and the specific heat capacity value was found by the following formula:

$$C = K_T / m_0 (\tau_T \tau_T^0),$$

where C = specific heat capacity, J/kg*K; m_0 = sample mass, g; K_T = instrument constant, which depends on the temperature and is found by the instrument calibration; τ_T^0 = temperature delay time with empty ampoule, s; τ_T = temperature delay time on heat meter, s.

The λ-calorimeter IT-λ-400 based on monotonous heating mode was used to measure heat conductivity. In the experiment, the readings of $n_0$temperature difference by plate thickness were measured with a stopwatch by the following formula to define the heat resistance:

$$R_s = ((1 + \sigma_c) S_{n0} / K_T * n_T) \ R_K,$$

where R_s = sample heat resistance, m^2*K/W; S = sample area, м^2; σ_c = $C_0 / 2(C_0 + C_c)$ allowance, taking into account the sample heat capacity C_0 (C_0 = total heat capacity of the sample tested, J/(kg*K); found before with the instrument IT-s-400; C_c, K_T, R_K = constant of the measuring instrument. The time was measured with a stopwatch with the accuracy of 0.01.s.

4. Determination of Relative and Absolute Viscosity of Suspensions: Relative viscosity of the modified liquid glass was found with viscometer VZ-246 [5]. The viscometer was placed in the rack and fixed horizontally with the balance. The vessel was placed under the viscometer nozzle. The nozzle hole was closed with a finger. Ihe material to be tested was poured into the viscometer to excess in order to have convex meniscus above the upper edge of the viscometer. The viscometer was filled slowly to avoid air bubbles. Then the nozzle hole was opened and as soon as the tested material appeared, the stopwatch was started and then stopped to calculate the discharge time. The values of relative viscosity were translated into the kinematic based on Russian Standard 8420-74.

The absolute (dynamic) viscosity of liquid glass was measured with a falling sphere viscometer (Geppler viscometer) intended for precise measurement of transparent Newton liquids and gases. In accordance with Geppler principle, the liquid viscosity is proportional to the falling time being measured. The falling time of a sphere rolling down inside the inclined pipe filled with the liquid being tested was measured. The time required for the sphere to cover the distance between the upper and lower ring marks on the pipe with the sample was measured with a stopwatch with the resolution 0.01 s. Further, the values of dynamic viscosity were calculated.

5. IR Spectrometry Technique: IR Fourier spectrometer (FSM) 1201 was used to obtain IR spectra of suspensions based on liquid glass and nanocomposites. The spectra were taken in the range of wave numbers 3994500 cm^{-1}.

3.3 RESULTS

3.3.1 Preparation of Fine Suspension
According to the reference sources, the most advantageous and urgent technique to introduce nanostructures into material is the application of fine suspensions of nanoproducts. They provide the fine modifier uniform distribution through the volume of the material modified.

The necessary concentration of nanostructures in suspension was chosen based on the basic binder mass and varied from 0.003 up to 0.3%. The suspensions were prepared in mechanical mortar mixing nanocomposites with liquid glass following the pre-selected mode with further processing in ultrasound bath (Figure 3) to prevent the nanostructure coagulation. The metal in nanostructures will possibly interact with liquid glass forming insoluble silicates. Due to this, the hydrosilicicacid is isolated from the liquid glass as an insoluble gel that results in compressing the liquid glass structure and increasing its moisture stability.

3.3.2 Analysis of Compositions on Liquid Glass Basis
1. Ultrasound Processing: According to the reference sources, the ultrasound technique is an effective dispergation method. Such, techniques are widely applied today. However, the lengthy soaking in ultrasound baths can result in partial or complete decomposition of nanostructures and their re-coagulation. The decomposition of nanostructures due to lengthy ultrasound action is conditioned by high temperature and pressure during the cavitations. The coagulation can be caused with the decomposition of solvate shell on disperse phase particles. It is necessary to define the processing time interval during which the optical density will be at the maximum, that is, it will correspond to the maximal saturation of nanocomposite suspension.

To define the time interval of ultrasound processing, four suspensions were prepared for each soaking period 3, 5, 10, and 15 min, respectively, and also the reference solution without the US processing for comparison. The concentration in all solutions was 0.003%.

Thus, the optimal time interval for ultrasound processing is 5 min. Further processing of suspensions for investigation will be carried out within this interval.

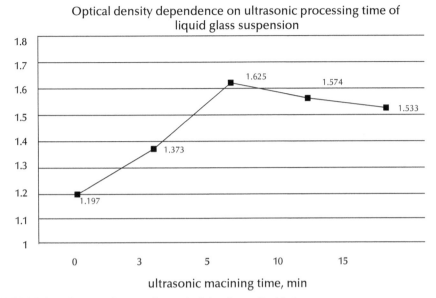

FIGURE 3 Diagram of suspension optical density on liquid glass.

2. Spectrophotometric Investigations: In accordance with the results of spectro-photometric investigations of modified films in the glue sample with Fe and glue with Ni, the shift of optical density is observed at some wavelengths, indicating the changes taking place when nanostructures are introduced (Figure 4).

At other wavelength values, the optical density only increases indicating the possibility to apply nanostructures as coloring pigments in paints. At the same time, the increase in the optical density of films with nanostructures in comparison with liquid glass films possibly indicates the increase in the density of compositions and formation of new structural elements in them.

3. Investigation of Heat physical Properties: To obtain the details of changes in heat physical characteristics, the heat capacity and thermal conductivity of the samples based on cardboard and modified liquid glass were investigated. The samples were prepared gluing several layers of cardboard with liquid glass modified with nanostructures. The sample dimensions were found by the technique for measuring heat capacity and thermal conductivity. The sample being tested was 15 mm in diameter and 10 mm in height. The sample dimensions were measured with micrometer with 0.01 mm accuracy. The sample mass was measured with the allowance not exceeding 0.001 g. Heat capacity of the samples was investigated on c-calorimeter IT-s-400. To find the thermal conductivity, the calorimeter IT-λ-400 was used. Further, the specific thermal conductivity of the sample was calculated as follows:

$$\lambda = h/R_s$$

where λ = specific thermal conductivity, W/m*K; h = sample thickness, m. The results of heat physical investigations are given in Table 1.

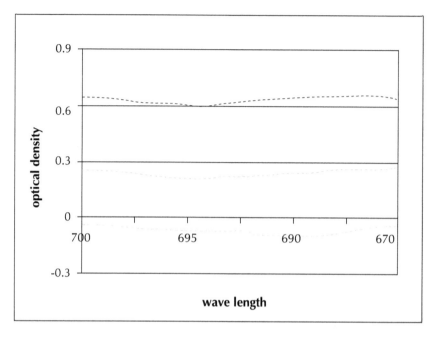

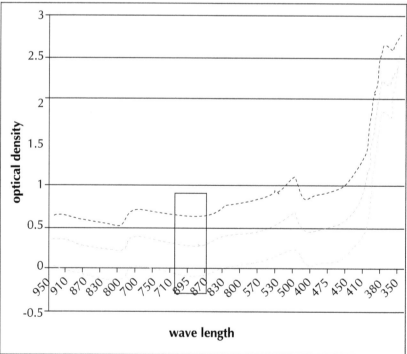

FIGURE 4   Shift of optical density for Fe in relation to Ni and liquid glass 1—Fe/C NC and liquid glass, 2—Ni/C NC and liquid glass, 3—liquid glass).

TABLE 1   Heat physical characteristics of the samples.

Samples	Cardboard /glue	Cardboard /glue with Fe (change in %)	Cardboard/glue with Ni (change in %)
Density, kg/m³	624.5	744 (↑19%)	669 (↑7%)
Heat capacity C_{spec}, J/kg*K	1790	2156 (↑20%)	2972 (↑66%)
Thermal conductivity λ, W/m*K	0.083	0.061 (↓27%)	0.064 (↓23%)

The heat capacity of the sample with Fe increased by 20% in comparison with the reference sample with Ni by 66%. At the same time, thermal conductivity decreased by 27 and 23% for the samples with Fe and Ni, respectively. So when nanostructures are introduced, in the average the characteristics change as follows: density increases by 13%, heat capacity by 40%, and thermal conductivity decreases by 25%. Further, using the experimental results, the temperature conductivity is calculated by the following formula:

$$a = \lambda/c\rho,$$

where a = temperature conductivity coefficient, λ = heat conductivity coefficient, c = heat capacity, ρ = density.

Inserting the experimental data into the formula, we can define the temperature conductivity values (in percent) as follows:

$$a = \lambda/c\rho = 0.75\lambda_0/1,4c_0*1.13\rho_0 = 0.47a_0$$

Or calculate separately as in the following:

$$a_1/a_0 = \lambda_1\rho_0C_0/\lambda_0\rho_1C_1 = 0.51$$
$$a_2/a_0 = \lambda_2\rho_0C_0/\lambda_0\rho_2C_2 = 0.43$$

where a_1, λ_1, ρ_1,C_1 = characteristics of the sample with Fe/C NC,
  a_2, λ_2, ρ_2,C_2 = characteristics of the sample with Ni/C NC,
  a_0, λ_0, ρ_0,C_0 = characteristics of non-modified sample.

Thus, temperature conductivity decreased by nearly 50% in comparison with the initial values (a_0).

When nanostructures are introduced, self-organization takes place. Nanoparticles structure the silicate matrix leading to the formation of new elements in the structure, thus increasing the material density and influencing its heat physical characteristics. When additional structural elements and new bonds are formed, the system internal energy increases, leading to heat capacity elevation and, consequently, temperature conductivity decreases. Thermal conductivity decrease of silicate paints when applied as a coating allows improving heat physical characteristics of the whole protective structure of a building. In turn, temperature conductivity decrease results in decreasing the amount of heat passing through the coating, thus preserving adhesive characteristics of the coating for long time.

4. IR Spectroscopy: To determine the interactions between nanostructures and liquid glass, the IR spectroscopy investigations were carried out as well. The spectra were taken in relation to water as the suspensions contained nanocomposite solution with liquid glass and water (Figure 5). The spectra were read with the reference tables. Water spectra demonstrate the region 27503750 cm^{-1} connected with O–H bond oscillations. They were observed in other regions as indicated by the values of wave numbers: 2380, 3850, and 3840 cm^{-1}. At the same time, the values 3,200 and 3,500 cm^{-1} appropriate for valence bound OH appeared in the spectra of nanostructures on Fe basis. In the region 1,6001,650 cm^{-1}, the bands of deformation oscillations of OH-groups were observed. Bands appropriate to the oscillations of Si-O-Me and Si-O-bonds (1,100400 cm^{-1}) [5] were seen in low frequency region.

The IR spectra contained wave numbers reflecting the interaction between metal/carbon nanocomposites and liquid glass, for example, peaks with wave numbers 407420 cm^{-1} (Si-Me). The spectra of suspensions vividly demonstrated peaks at 1,020 and 1,100 cm^{-1}, appropriate to the oscillations of Si-O-Si and Si-O-C. However, in the spectrum of liquid glass containing Fe/C nanocomposites those values were shifted from 1,104 to 1,122 cm^{-1}, there was also the shift from 600 to 660 cm^{-1} caused by the change in oscillations of some bonds under the nanocomposite action. The spectrum obtained resembles the liquid glass spectrum from reference sources [5].

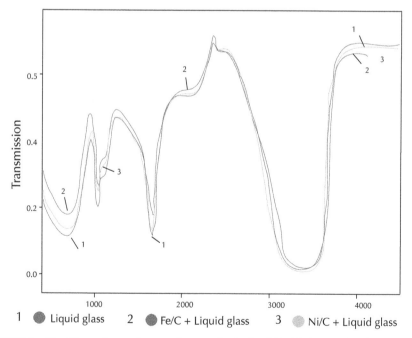

FIGURE 5   The IR spectrum of suspensions on the basis of liquid glass and various types of nanocomposites (1liquid glass, 2Fe/C NC and liquid glass, 3Ni/C NC and liquid glass).

The spectra demonstrates the absence of strong interactions, mostly the intensity changes, thus indicating the formation of a large number of definite bonds, for example, Si-O showing the structuring of liquid medium.

For suspensions on liquid glass basis, the relative and absolute viscosities were found. Taking into account the aforementioned data for defining the optimal ultrasound processing time, the suspensions for viscosity test were prepared with ultrasound pre-processing within 5 min.

5. Determination of Relative and Absolute Viscosity of Suspensions: Dynamic viscosity is the relation between force unit required to shift the liquid layer for distance unit and layer area unit. In metric system, it is given as dyne-second per square centimeter called "poise" as in the following:

$$\rho_1 - \rho_2,$$

where η = dynamic viscosity [mPa·s], τ = sphere movement time [s], ρ_1 = sphere density according to the test certificate [g/cm^3], ρ_2 = sample density, [g/cm^3], κ = sphere constant according to the test certificate [mPa·cm^3/g], F = work angle constant.

The experiment was carried out with pipe inclination for the sample 80° = >F = 1.0

$$\rho_1 = 15.2 \text{ g/cm}^3$$

$$\rho_2 = 1.45 \text{ g/cm}^3$$

$$\kappa = 0.7$$

Liquid glass viscosity: τ_{cp} = 9.60 s

$$\eta = 9.6(15.2 - 1.45) \cdot 0.7 \cdot 1 = 92.4 \text{ Pa·s}$$

Viscosity of liquid glass modified with iron containing nanocomposite:

$$\eta = 6.97(15.2 - 1.45) \cdot 0.7 \cdot 1 = 67.09 \text{ Pa·s}$$

Viscosity of modified liquid glass with nickel containing nanocomposite:

$$\eta = 8.76(15.2 - 1.45) \cdot 0.7 \cdot 1 = 84.32 \text{ Pa·s}$$

Kinematic viscosity is the ratio between dynamic viscosity and liquid density.

Based on the data obtained when measuring the viscosity of liquid glass and fine suspension with Fe/C and Ni/C NC, the diagrams were constructed (Figures 6, 7).

Having analyzed this diagram, we can conclude that when a nanocomposite is introduced, the viscosity decreases. Such, phenomenon will have a positive effect in the process of silicate material production and application. For example, better application in silicate paint, much easier foaming process in foam glass.

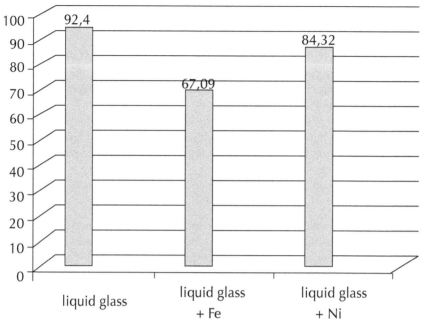

FIGURE 6   Dynamic viscosity.

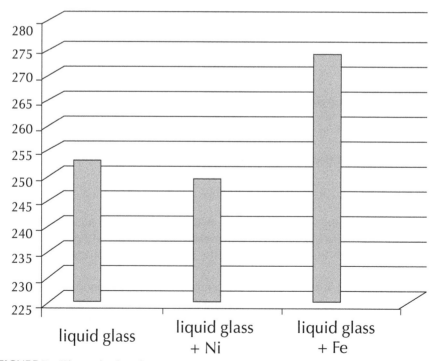

FIGURE 7   Kinematic viscosity.

In this diagram, we can see that nanostructures work differently. The viscosity of fine suspension with Ni/C NC is less than kinematic viscosity with Fe/C NC. This can be caused by coagulation, aggregation of Fe/C NC.

TABLE 2 Characteristics of kinematic and dynamic viscosity of liquid glass and its analogs modified with Ni/C and Fe/C nanocomposites.

Sample	Kinematic viscosity, mm²·s⁻¹	Dynamic viscosity, Pa·s (change in %)
Liquid glass	253.3	92.4
Liquid glass with Ni/C nanocomposite	244.3 (↓4%)	84.32 (↓9%)
Liquid glass with Fe/C nanocomposite	275.1 (↑9%)	67.09 (↓27%)

The viscosity of compositions on liquid glass basis depends on nanocomposite composition: in some cases, the viscosity increase is observed but in other cases the decrease is observed. The viscosity increase is connected with the elevation of intermolecular friction that can be caused by the influence of nanoparticles on liquid glass structuring and density increase. The decrease in suspension viscosity can be explained by the fact that NC is surrounded by liquid glass molecules resulting in the decrease in their interactions and intermolecular friction.

3.4 CONCLUSION

Based on the results obtained, it can be concluded that nanostructures have a positive effect on liquid glass properties.

The optimal time interval of suspension ultrasound processing was found experimentally. It equals 5 min and corresponds to the maximum solution saturation with nanocomposites.

The data of optical density investigation of liquid glass films and IR spectra of liquid glass solutions indicate the changes in material structure when introducing metal/carbon nanocomposites.

The changes in heat physical and viscous properties of suspensions on liquid glass indicate the self-organization processes when introducing nanostructures. At the same time, the results differ when introducing nanocomposites of different compositions.

The application of nanocomposites will allow increasing the life of paints and consequently, their storage, and covering ability. Also not only potassium but also sodium liquid glass can be used, since potassium liquid glass is significantly more expensive than sodium one and its manufacturing is limited. To improve the characteristics, only small concentrations of nanoproduct are required, which will positively affect the material production cost.

KEYWORDS

- **Dynamic viscosity**
- **Globular film nanostructures**
- **Nanotechnology**
- **Nanoreactors**
- **Thermal conductivity**

REFERENCES

1. Kodolov, V. I. and Khokhriakov, N. V. Chemical physics of formation and transformation processes of nanostrcutures and nanosystems.Izhevsk. *Izhevsk State Agricultural Academy*, **1, 2** 365, 415 (2009).
2. Kodolov V. I., Vasilchenko, Y. M., Akhmetshina, L. F., Shklyaeva, D. A., Trineeva, V. V., Sharipova, A. G., Volkova, E. G., Ulyanov, A. L., and Kovyazina, O. A.; declared on 17.10.2008, published on 27.06.2010.
3. Kodolov V. I., Kodolova V. V. (Trineeva), Semakina, N. V., Yakovlev, G. I., and Volkova, E. G. et al; declared on 28.08.2006, published on 27.10.2008.
4. Platunov, E. S., Buravoy, S. E., Kurepin, V. V., and Petrov, G. S. *Heat-physical measurements and instruments*. E.S. Platunov (Ed.) L., Mashinostroenie (1986).
5. Plyusnina, I. I. *IR spectra of silicates*. M.: Publishing House of Moscow State University, p. 190 (1967).

4 Updates on Polymeric Nanocomposites Reinforcement

G. V. Kozlov, Yu. G. Yanovskii, B. Zh. Dzhangurazov, and G. E. Zaikov

CONTENTS

4.1 INTRODUCTION

A filler (nanofiller) is introduced in polymers quite often with the purpose of increasing the stiffness of the latter. This effect is called polymer composites (nanocomposites) reinforcement and it is characterized by reinforcement degree E_c/E_m (E_n/E_m), where E_c, E_n, and E_m are elasticity module of composite, nanocomposite, and matrix polymer, respectively. The indicated effect significance results into a large number of quantitative model developments, describing reinforcement degree: Micromechanical [1], percolation [2], and fractal [3] ones. The principal distinction of the indicated modes is the circumstance that the first ones take into consideration filler (nanofiller) elasticity modulus and the last two do not. The percolation [2] and fractal [3] models of reinforcement assume that filler (nanofiller) role comes to modification and fixation of matrix polymer structure. Such approach is obvious enough, if the difference of elasticity modulus of filler (nanofiller) and matrix polymer are to be taken into consideration. So, in this chapter of nanocomposites, low density polyethylene/Na$^+$-montmorillonite, the matrix polymer elasticity modulus makes up to 0.2 GPa [4] and nanofiller 400420 GPa [5], that is the difference makes up more than three orders. It is obvious that at such conditions, organoclay strain is practically equal to zero and nanocomposite behavior in mechanical tests is defined by polymer matrix behavior.

Lately, it was offered to consider polymers' amorphous state structure as natural nanocomposites [6]. Within the frameworks of cluster model of polymers' amorphous state structure, it is supposed that the indicated structure consists of local order domains (clusters), immersed in loosely packed matrix, in which the entire polymer free volume is concentrated [7, 8]. The clusters consist of several collinear densely packed statistical segments of different macromolecules, that is they are amorphous analog of crystallites with stretched chains. It has been shown that the clusters are nanoworld objects (true nanoparticles–nanoclusters) [9] and in case of polymers representation as natural nanocomposites, they play the roles of nanofillers and loosely packed matrix—nanocomposite matrix. It is significant to note that the nanoclusters' dimensional effect is identical to the indicated effect for particulate filler in polymer nanocomposites—sizes decrease of both nanoclusters [10] and disperse particles results into sharp enhancement of nanocomposites reinforcement degree (elasticity modulus) [11]. In connection with the indicated observations, the question arises: How nanofiller introduction in polymer matrix influences on nanoclusters' size and how the variation of the latter influences on nanocomposite elasticity modulus value. The purpose of the present chapter is these two problems solution on the example of two classes' nanocomposites: Polymer/organoclay (linear low density polyethylene/Na$^+$-montmorillonite [4]) and particulate-filled nanocomposite (low density polyethylene/calcium carbonate [12]).

4.2   EXPERIMENTAL

In case of nanocomposites polymer/organoclay, linear low density polyethylene (LLDPE) of mark Dewlex-2032, having melt fluidity index 2.0 g/10 min and density 926 kg/m^3 that corresponds to crystallinity degree 0.49 was used as matrix polymer. Modified Na$^+$-montmorillonite (MMT), obtained by cation exchange reaction between MMT and quaternary ammonium ions, was used as nanofiller. The MMT contents make up 17 mass % [4].

Nanocomposites linear low density polyethylene/Na$^+$-montmorillonite (LLDPE/MMT) was prepared by components blending in melt using Haake twin screw extruder at temperature 473K [4].

Tensile specimens were prepared by injection molding on Arburg Allounder 305-210-700 molding machine at temperature 463K and pressure 35 MPa. Tensile tests were performed by using tester Instron of model 1,137 with direct digital data acquisition at temperature 293K and strain rate ~ 3.35×10^{-3} s^{-1}. The average error of elasticity modulus determination makes up 7%, yield stress 2% [4].

In case of particulate-filled nanocomposites, low density polyethylene (LDPE) of mark 10803-020 with melt fluidity index 0.6 g/10 min was used as matrix polymer and nanodimensional calcium carbonate (CaCO$_3$) in the form of compound mark Nano-Cal NC-KO117 (China) with particles size of 80 nm and mass contents blending in melt on twin screw extruder Thermo Haake model, Reomex RTW 25/42, and production of German Federal Republic. Blending was performed at temperature 448463K and screw speed 1525 rpm during 5 min. Testing samples were obtained by casting under pressure method on casting machine Test Samples Molding Apparate RR/TS MP of firm Ray-Ran (Taiwan) at temperature 473K and pressure 8 MPa [12].

Uniaxial tension mechanical tests were performed on samples in the form of two-side spade with sizes according to GOST 11262-80. Tests were conducted on universal testing apparatus Gotech Testing Machine CT-TCS 2000, production of German Federal Republic at temperature of 293K strain rate of $\sim 2 \times 10^{-3}$ s^{-1} [12].

4.3 RESULTS

For the solution, first from the indicated problems, the statistical segments number in one nanocluster n_{cl} and its variation at nanofiller contents change should be estimated. The parameter n_{cl} calculation consistency includes the following stages. First, the nanocomposite structure fractal dimension d_f is calculated according to the following equation [13]:

$$d_f = (d-1)(1+\nu) \tag{1}$$

where d is dimension of Euclidean space, in which a fractal is considered (it is obvious that in our case $d = 3$), n is Poisson's ratio, which is estimated according to mechanical tests results with the aid of the relationship as in the following [14]:

$$\frac{\sigma_Y}{E_n} = \frac{1-2\nu}{6(1+\nu)}, \tag{2}$$

where σ_Y and E_n are yield stress and elasticity modulus of nanocomposite, respectively.

Then nanoclusters relative fraction φ_{cl} can be calculated by using the following equation [8]:

$$d_f = 3 - 6\left(\frac{\varphi_{cl}}{C_\infty S}\right)^{1/2} \tag{3}$$

where C_∞ is characteristic ratio, which is a polymer chain statistical flexibility indicator [15] and S is macromolecule cross-sectional area.

The value C_∞ is a function of d_f according to the relationship as in the following [8]:

$$C_\infty = \frac{2d_f}{d(d-1)(d-d_f)} + \frac{4}{3}. \tag{4}$$

The value S for low density polyethylene is accepted equal to 14.9Å^2 [16]. Macromolecular entanglements cluster network density ν_{cl} can be estimated as follows [8]:

$$\nu_{cl} = \frac{\varphi_{cl}}{C_\infty l_0 S} \tag{5}$$

where l_0 is the main chain skeletal bond length, which for polyethylenes is equal to 0.154 nm [17].

Then the molecular weight of chain part between nanoclusters M_{cl} was determined according to the following equation [8]:

$$M_{cl} = \frac{\rho_p N_A}{\nu_{cl}}$$

(6)

where ρ_p is polymer density, which for the studied polyethylenes is equal to ~ 930 kg/ m^3, N_A is Avogadro number.

And at last, the value n_{cl} is determined as follows [8]:

$$n_{cl} = \frac{2M_e}{M_{cl}}$$

(7)

where M_e is molecular weight of chain part between entanglements traditional nodes ("binary hooking" [18]), which is equal to 1,390 g/mol for low density polyethylenes [18].

In Figures 1 and 2, the dependence of elasticity modulus E_n on value n_{cl} for nano-composites LLDPE/MMT and LDPE/CaCO$_3$ are adduced, accordingly. As one can see, for both studied nanocomposites E_n enhancement at n_{cl} decreasing is observed. Such behavior of both classes polymer nanocomposite is completely identical to the dependence of E_n on nanofiller particles diameter for particulate-filled nanocomposites [11] and the dependence $E(n_{cl})$ for natural nanocomposite [10].

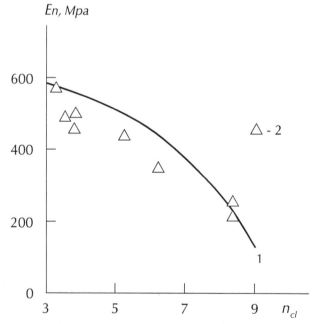

FIGURE 1   The dependence of elasticity modulus E_n on statistical segments number per one nanocluster n_{cl} for nanocomposites LLDPE/MMT. (1) Calculation according to the Equation (14), (2) the experimental data.

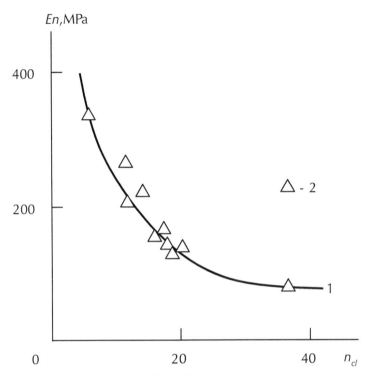

FIGURE 2   The dependence of elasticity modulus En on statistical segments number per one nanocluster ncl for nanocomposites LDPE/CaCO$_3$. (1) calculation according to the Equation (14), (2) the experimental data.

Let us consider the physical grounds of n_{cl} reduction at nanofiller mass contents W_n growth. For this purpose, several methods can be used and the first from them is applied to nanocomposites LLDPE/MMT. The main equation of polymer nanocomposites reinforcement percolation model is in the following [9]:

$$\frac{E_n}{E_m} = 1 + 11\left(\varphi_n + \varphi_{if}\right)^{1.7}$$ (8)

where φ_n and φ_{if} are relative volume fractions of nanofiller and interfacial regions, respectively.

The value φ_n can be determined according to the following equation [5]:

$$\varphi_n = \frac{W_n}{\rho_n}$$ (9)

where ρ_n is nanofiller density, which is equal to ~1700 kg/m^2 for Na$^+$-montmorillonite [5].

Further, the Equation (8) allows to estimate the value φ_{if}. In Figure 3, the dependence $n_{cl}(\varphi_{if})$ for nanocomposites LLDPE/MMT is adduced. As one can see n_{cl} reduction at φ_{if} growth is observed, that is formed on organoclay surface densely packed (and, possible, subjecting to epitaxial crystallization [9]) interfacial regions as though are "pull apart" nanoclusters, decreasing in them statistical segments number.

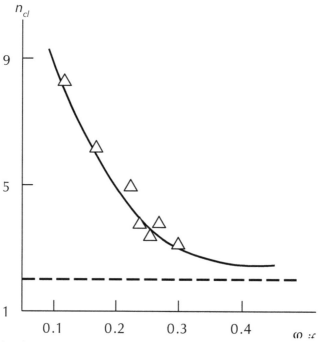

FIGURE 3   The dependence of statistical segments number per one nanocluster n_{cl} on interfacial regions relative fraction φ_{if} for nanocomposites LLDPE/MMT. Horizontal shaded line indicates the minimum value $n_{cl} = 2$.

Another method can be used for similar effect explanation in nanocomposites LDPE/CaCO$_3$. It is known [19] that the densely packed region's maximum fraction φ_{dens} for polymeric materials is given by the following percolation relationship:

$$\varphi_{dens} = \left(\frac{T_m - T}{T_m}\right)^{\beta_T} \tag{10}$$

where T_m and T are melting and testing temperatures, respectively (for low density polyethylenes $T_m = 398K$ [20]), β_T is a thermal cluster order parameter, which is equal to 0.55 for polymers [21].

It is quite obvious that matrix polymer space filled by nanofiller decreases polymer matrix fraction in nanocomposite structure, particularly the latter contents at large (up to 50 mass %). Therefore, the reduced value φ_{dens} (φ_{dens}^{red}) should be used, which is determined as follows [9]:

$$\varphi_{dens}^{red} = \frac{\varphi_{dens}}{1-\varphi_n} \tag{11}$$

where value φ_n can be calculated according to the Equation (9) and nanofiller density ρ_n is estimated according to the following formula [9]:

$$\rho_n = 0.188\left(D_p\right)^{1/3} \tag{12}$$

where D_p is diameter of disperse nanofiller particles.

For nanocomposites LDPE/CaCO$_3$ as structure densely packed regions, the equation should be accepted as follows [9]:

$$\varphi_{dens} = \left(1-K\right)\varphi_{cl} + \varphi_{if} \tag{13}$$

where K is crystallinity degree, which for low density polyethylenes can be accepted equal to ~0.50 [20].

In Figure 4, the dependence φ_{dens}^{red} (W_n) for nanocomposites LDPE/CaCO$_3$ is adduced. As one can see, the values φ_{dens}^{red} determined according to the Equations (10), (11), and (13), agree well with one another. The Equation (11) shows the cause of n_{cl} reduction at W_n (or φ_n) growth, that is the decrease in polymer matrix fraction, accompanied by φ_{cl} reduction results into v_{cl} decreasing, according to the Equation (5); M_{cl} increasing, according to the formula (6); and n_{cl} reduction, according to the Equation (7).

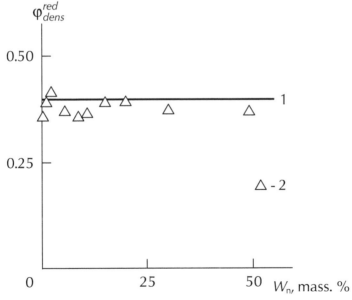

FIGURE 4 The dependence of reduced relative fraction of densely packed regions φ_{dens}^{red} on nanofiller mass contents W_n for nanocomposites LDPE/CaCO$_3$. Calculation: (1) according to the Equations (10) and (11); (2) according to the Equation (13).

In this chapter [22], the theoretical dependence of E_n as a function of cluster model parameters for natural nanocomposites was obtained as in the following:

$$E_n = c\left(\frac{\varphi_{cl} v_{cl}}{n_{cl}}\right) \qquad (14)$$

where c is constant, accepting equal to 5.9′ 10^{-26} m³ for LLDPE and 1.15 ′ 10^{-26}m³ for LDPE.

In Figures 1 and 2, the theoretical dependences $E_n(n_{cl})$ calculated according to the Equation (14), for nanocomposites LLDPE/MMT and LDPE/CaCO₃, respectively, are adduced. As one can see, the calculation according to the Equation (14) showed a good enough correspondence to the experiment (their average discrepancy makes up ~12%, which is comparable with the mechanical tests experimental error). Therefore, at organoclay mass contents W_n increasing within the range of 07 mass %, n_{cl} value reduces from 8.40 up to 3.17, which is accompanied by nanocomposites LLDPE/MMT elasticity modulus growth from 206 up to 569 MPa; and at mass contents of CaCO₃ increasing within the range of 050 mass %, n_{cl} value reduces from 36.0 up to 5.2 that results into nanocomposites LDPE/CaCO₃ elasticity modulus growth from 85 up to 340 MPa.

Let us note that constant c in the Equation (14) is function of matrix polymer characteristics, which was expected. In Figure 5, the dependence of constant c on matrix polymer elasticity modulus E_m for LLDPE, LDPE, polypropylene (PP), and polycarbonate (PC) [22] is adduced. As one can see, the value c grows at E_m increasing and is described by the following empirical Equation:

$$c = 9.71 \times 10^{-26} E_m \qquad (15)$$

where E_m is given in GPa.

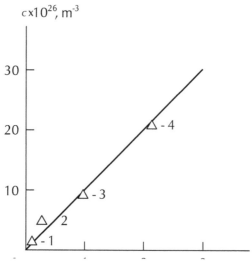

FIGURE 5   The dependence of constant c in the Equation (14) on matrix polymer elasticity modulus E_m for LDPE (1), LLDPE (2), PP (3), and PC (4).

Following the plots $E_n(n_{cl})$ of Figures 1 and 2, a very important aspect of comparison should be noted. As one can see, they have different shapes for the studied nanocomposites: For LLDPE/MMT—a convex one and for LDPE/CaCO$_3$—a concave one. The authors [9] qualitatively obtained similar theoretical dependences of reinforcement degree E_n/E_m on nanofiller mass contents W_n for nanocomposites polymer/organoclay and particulate-filled polymer nanocomposites. For the value E_n/E_m estimation in the chapter [9], the following equations are used:

$$\frac{E_n}{E_m} = 1 + 0.32 W_n^{1/2} l_{st}, \qquad (16)$$

and particulate-filled polymer nanocomposites

$$\frac{E_n}{E_m} = 1 + \frac{0.19 W_n l_{st}}{D_p^{1/2}} \qquad (17)$$

In the Equations (16) and (17), l_{st} is statistical segment length W_n is given in mass %, l_{st}, and D_p in nm. The authors [9] explained a qualitatively different shape of curves $E_n/E_m(W_n)$ as follows. It is known that the layered silicates at contents W_n ^{3} 10 mass % display clearly expressed the tendency to separate platelets aggregation with "packents" (tactoids) formation, which reduces noticeably their reinforcement degree [5]. Since the Equations (16) and (17) are empirical ones, obtained on the basis of experimental data analysis, they take into consideration this effect by the dependence of E_n/E_m on W_n to the half power for the layered nanofillers (the Equation (16)) and to the first power for disperse particles (the Equation (17)).

The data of the present chapter allow interpreting this aspect from a new point of view. The theory and experiment correspondence, which follows from the data of Figures 1 and 2, is obtained at the usage in the Equation (14) as v_{cl} actually the value v_{cl} (the Equation (5)) for nanocomposites LLDPE/MMT and the value v_{dens} (the Equation (13)) for nanocomposites LDPE/CaCO$_3$. In other words, the interfacial regions' role for the indicated nanocomposites is essentially distinguished: If the indicated regions form macromolecular entanglements cluster network in case of LDPE/CaCO$_3$ then for the nanocomposites LLDPE/MMT, this effect is absent. One ought to suppose that such difference is due to the interfacial regions arrangement in case of nanocomposites polymer/organoclay in "galleries" between exfoliated (intercalated) silicate platelets that strongly restricts their coupling with nanocomposite bulk polymer matrix [9]. Nevertheless, for both indicated classes of nanocomposites, interfacial regions are their structure reinforcement element (Equation (8)) or solid body component in reinforcement percolation models [2].

And let us consider the practical aspect of organoclay and disperse particles usage as nanofiller for polymer nanocomposites. In Figure 6, the dependence of reinforcement degree on nanofiller mass contents W_n are adduced for nanocomposites LLDPE/MMT and LDPE/CaCO$_3$, which are obtained experimentally. As it follows from these dependence comparison, at relatively small W_n (<30 mass %) nanocomposites LLDPE/MMT have higher values E_n/E_m at the same W_n, but at W_n enhancement the situation changes on the contrary. Additionally, one should take into account that paeticulate-filled nanocomposites are more technological in the production process and have smaller cost. As it has been noted above, organoclay exfoliation at its content more than 10 mass % is very difficult (in any case, modern technologies do not give such possibility) and therefore, the dependence $E_n/E_m(W_n)$ for nanocomposites polymer/organoclay reaches rapidly asymptotic branch, as shown in Figure 6. Let us note that nanofiller disperse particles aggregation suppression can give the larger effect, namely, the displacement of the curve 2 in Figure 6 to the left, the more so there exists a number of modes at present [9, 11, and 23].

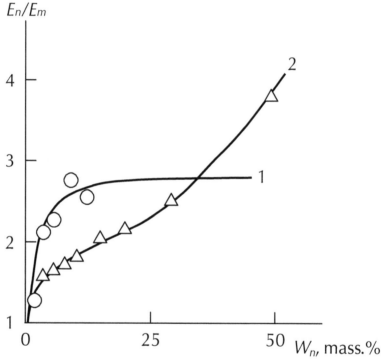

FIGURE 6   The dependence of reinforcement degree E_n/E_m on nanofiller mass contents W_n for nanocomposites LLDPE/MMT (1) and LDPE/CaCO$_3$ (2).

4.4   CONCLUSION

Hence, the results obtained in the present chapter demonstrated common reinforcement mechanism of natural and artificial (filled with inorganic nanofiller) polymer nanocomposites. The statistical segments number per one nanocluster reduction at nanofiller contents growth is such a mechanism on segmental level. The observed effect physical foundation is densely packed interfacial regions formation in artificial nanocomposites. The structural causes of different shape of the dependences of elasticity modulus on nanofiller contents for polymer nanocomposites of various classes have been considered.

KEYWORDS

- **Clusters**
- **Nanocomposites**
- **Nanofiller**
- **Nanofiller density**
- **Polymer nanocomposites**
- **Polymeric materials**

REFERENCES

1. Ahmed, S. and Jones, F. R. *J. Mater. Sci.*, **25**(12), 4933–4942 (1990).
2. Bobryshev, A. N., Kozomazov, V. N., Babin, L. O., and Solomatov, V. I. *Synergetics of Composite Materials.*, NPO ORIUS, Lipetsk, p. 154 (1994).
3. Kozlov, G. V., Yanovskii, Yu. G., and Karnet, Yu. N. *Structure and Properties of Particulate-Filled Polymer Composites: Fractal Analysis.*, Al'yanstransatom, Moscow, p. 363 (2008).
4. Hotta, S. and Paul, D. R. *Polymer*, **45**(21), 7639–7654 (2004).
5. Sheng, N., Boyce, M. C., Parks, D. M., Rutledge, G. C., Abes, J. I., and Cohen, R. E. *Polymer*, **45**(2), 487–506 (2004).
6. Bashorov, M. T., Kozlov, G. V., Mikitaev, A. K. *Nanostructures and Properties of Amorphous Glassy Polymers.* Publishers of D. I. Mendeleev RKhTU, Moscow, p. 269 (2010).
7. Kozlov, G. V. and Novikov, V. U. *Uspekhi Fizicheskikh Nauk*, **171**(7), 717–764 (2001).
8. Kozlov, G. V. and Zaikov, G. E. *Structure of the Polymer Amorphous State.* Brill Academic Publishers, Utrecht, Boston, p. 465 (2004).
9. Mikitaev, A. K., Kozlov, G. V., and Zaikov, G. E. *Polymer Nanocomposites: Variety of Structural Forms and Applications.* Nova Science Publishers, Inc., New York, p. 319 (2008).
10. Amirshikhova, Z. M., Bashorov, M. T., Kozlov, G. V., and Magomedov, G. M. Mater. Of VIII International Sci.-Pract. conf. "Materialy i Tekhnologii XXI Veka". Penza, PSU, pp. 710 (2010).
11. Edwards, D. C. *J. Mater. Sci.*, **25**(12), 4175–4185 (1990).
12. Sultonov, N. Zh. and Mikitaev, A. K. Mater of VI International Sci. Pract. conf. "Novye Polimernye Ko,pozitsionnye Materialy". Nal'chik, KBSU, pp. 392–398 (2010).
13. Balankin, A. S. *Synergetics of Deformable Body.* Publishers of Ministry Defence SSSR, Moscow, p. 404 (1991).
14. Kozlov, G. V. and Sanditov, D.S. *Anharmonic Effects and Physical-Mechanical Properties of Polymers.*, Nauka, Novosibirsk, p. 261 (1994).
15. Budtov, V. P. *Physical Chemistry of Polymer Solutions.* Khimiya, Sankt-Peterburg, p. 384 (1992).
16. Aharoni, S. M. *Macromolecules*, **18**(12), 2624–2630 (1985).
17. Aharoni, S. M. *Macromolecules*, **16**(9), 1722–1728 (1983).
18. Wu, S. *J. Polymer Sci.: Part B: Polymer Phys.*, **27**(4), 723–741 (1989).
19. Family, F. *J. Stat. Phys.*, **36**(5/6), 881–896 (1984).
20. Kalinchev, E. L. and Sakovtseva, M. B. *Properties and Processing of Thermoplastics.* Khimiya, Leningrad, p. 288 (1983).
21. Kozlov, G. V., Gazaev, M. A., Novikov, V. U., and Mikitaev, A. K. *Pis'ma v ZhTF*, **22**(16), 3138 (1996).
22. Bashorov, M. T., Kozlov, G. V., Zaikov, G. E., and Mikitaev A. K. In Handbook of *Condensed Phase Chemistry*. G. E Zaikov and V. K. Kablov (Eds.). Nova Science Publishers, Inc., New York, pp. 3339 (2010).
23. Aphashagova, Z. Kh., Kozlov, G. V., Burya, A. I., and Mikitaev, A. K. *Materialovedenie*, (9), 1013 (2007).

5 Updates to Properties of Metal/Carbon Nanocomposite

M. A Chashkin, V. I. Kodolov, A. I. Zakharov,
Yu. M. Vasilchenko, M. A. Vakhrushina, and
V. V. Trineeva

CONTENTS

5.1 INTRODUCTION

This decade has been heralded by a large-scale replacement of conventional metal structures with structures from polymeric composite materials (PCM). Currently, we are facing the tendency of production growth of PCM with improved operational characteristics. In practice, PCM characteristics can be improved while applying modern manufacturing technologies, for example the application of "binary" technologies of prepreg production [1], as well as the synthesis of new polymeric PCM matrices, or modification of the existing polymeric matrices with different fillers.

The most cost-efficient way to improve operational characteristics is to modify the existing polymeric matrices; therefore, currently the group of polymeric materials modified with nanostructures (NS) is of special interest. The NS are able to influence the supermolecular structure, stimulate self-organization processes in polymeric matrices in super small quantities, thus contributing to the efficient formation of a new phase "medium modified—nanocomposite" and qualitative improvement of the characteristics of final product—PCM. This effect is especially visible when NS activity increases, which directly depends on the size of specific surface, shape of the particle, and its ultimate composition [2]. Metal ions in the NS used in this work also contribute to the activity increase as they stimulate the formation of new bonds.

The increase in the attraction force of two particles is directly proportional to the growth of their elongated surfaces; therefore, the possibility of NS coagulation and decrease in their efficiency as modifiers increases together with their activity growth. This fact and the fact that the effective concentrations to modify polymeric matrices are usually in the range below 0.01 mass % impose specific methods for introducing NS into the material modified. The most justified and widely applicable are the methods for introducing NS with the help of fine suspensions (FS). This introduction method allows most uniformly distributing particles in the volume of the medium modified, decreasing the possibility of their coagulation and preserving their activity during storage.

5.1.1 Background

In the process of quantum-chemical modeling, the fragments imitating the initial reagents are optimized: epoxy diane resin (EDR), polyethylene polyamine (PEPA), cobalt/carbon nanocomposite (Co/C NC), nickel/carbon nanocomposite (Ni/C NC), and copper/carbon nanocomposite (Cu/C NC) with the inclusion of Co^{2+}, Ni^{2+}, Cu^{2+} ions (Figure 1). For each of these fragments, the absolute value of binding energy is defined E_{EDR}, E_{PEPA}, $E_{NC(Me)}$, where Me – cobalt, nickel, or copper ion.

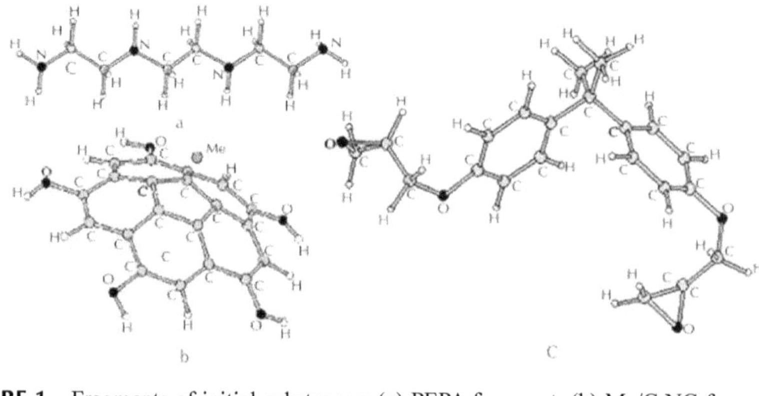

FIGURE 1 Fragments of initial substances: (a) PEPA fragment; (b) Me/C NC fragment; (c) EDR fragment.

As the modification process initially assumed the production of FS of NC on PEPA basis, the fragments imitating the behavior of the corresponding suspensions of Co/C, Ni/C, and Cu/C nanocomposites were optimized (Figure 2) and their absolute values of binding energy $E_{PEPA-NC(Co)}$, $E_{PEPA-NC(Ni)}$, and $E_{PEPA-NC(Cu)}$ were defined.

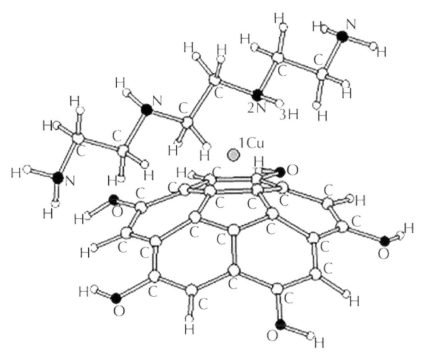

FIGURE 2   Fragment of Cu/C NC FS on PEPA basis.

The next step was to model the influence of FS of nanocomposite on epoxy resin (ER). The complexes formed similarly with the previous ones were optimized, and the absolute values of binding energy were found for each of them $E_{EDR-PEPA-NC(Co)}$, $E_{EDR-PEPA-NC(Ni)}$, and $E_{EDR-PEPA-NC(Cu)}$. The absolute values of binding energy are given in Table 1.

TABLE 1   Absolute binding energies of the fragments.

	$E_{NC(Me)}$, KJ/mol	$E_{PEPA-NC(Me)}$, KJ/mol	$E_{EDR-PEPA-NC(Me)}$, KJ/mol
Co²⁺	−19116.50	−29992.05	−51486.96
Ni²⁺	−18562.38	−29098.04	−50621.15
Cu²⁺	−18340.32	−28764.72	−50315.94
E_{EDR}, KJ/mol	−21424.25		
E_{PEPA}, KJ/mol	−10131.36		

Using the data from Table 1 by the following formula:

$$E_1 = E_{EDR-PEPA-NC(Me)} - E_2 - E_{EDR}$$
$$E_2 = E_{PEPA-NC(Me)-E_{PEPA}}$$

The relative interaction energies of molecular complexes E_1 are calculated and the diagram is arranged (Figure 3).

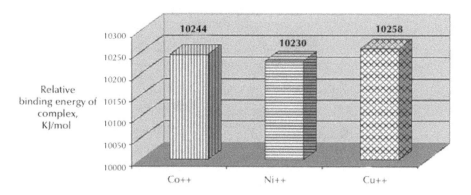

FIGURE 3 Diagram of relative interaction energies of molecular complexes.

From the diagram, it is seen that the relative interaction energy of molecular complex with Cu/C NC is higher in comparison with the complexes with Co/C NC and Ni/C NC content. As the polymer forms the strongest complexes with Cu/C NC, therefore after the modification, it will be the most effective one. The detailed analysis of the lengths of the bonds formed and effective charges before and after the optimization of fragments imitating PEPA interaction with Cu/C NC (Table 2, Table 3) indicates that stable coordination bonds were formed between NC fragment and PEPA (between copper ion and nitrogen atom of amine group NH of PEPA). It was found that after the interaction of two fragments studied, a part of electron density of N atom participating in the bond shifted to Cu atom, resulted in NH bond weakening.

TABLE 2 Bond lengths.

Bond designation	Bond length before optimization, Å	Bond length after optimization, Å
Cu-N	2.82	1.95
Cu-C	2.65	2.25

TABLE 3   Effective charges.

Atom number (see Figure 2)	Atom designation	Effective charge before optimization	Effective charge after optimization
1	Cu (copper)	0.138	−0.250
2	N (nitrogen)	−0.066	0.420
3	H (hydrogen)	0.045	0.027

The bond weakening is indirectly confirmed by the increase in the effective charge of H atom and slight change in the wave number in oscillatory spectra calculated. The wave number of NH bond before the optimization was 3,360 cm^{-1}, and after −3,354 cm^{-1}, which correlates with the data of IR spectra obtained with the help of IR Fourier spectrometer. For instance, in the spectrum of PEPA and NC suspension on PEPA basis with nanocomposite concentration 0.03%, the shift of wave numbers of peaks of amine groups is observed from 3,280 to 3,276 cm^{-1}.

The modeling of hardening process with the participation of ER, PEPA, and Cu/C nanocomposite is given in Figure 4(a) and Figure 4(b), respectively. From the geometry of optimized molecular systems, it is seen that the introduction of Cu/C NC into the system leads to its self-organization (Figure 4(b)) and formation of presumably coordination polymer. The formation of coordination polymer due to the introduction of nanocomposite active particles can result in increasing the adhesive strength and thermal stability as the total number of bonds in polymer grid grows and more energy will be required for its thermal destruction. At the same time, the formation of nanocomposite metal coordination bond with PEPA nitrogen can increase the stability of FS formed. Such coordination results in PEPA activity growth during the hardening of EDR.

Thus, quantum-chemical modeling allows predicting the interaction processes of components while hardening the ER with PEPA and active participation of copper/carbon nanocomposite.

5.2   EXPERIMENTAL

5.2.1   Materials

Currently, there is a huge need in modern epoxy systems with improved operational characteristics that can be reached, as previously mentioned, when modifying them with NS [3]. Therefore, in this work the modification processes of cold-hardened model epoxy composition containing epoxy Diane resin ED-20 State Standard (GOST) 10587-84 in the amount of 100 weight fractions and PEPA grade a Technical Condition (TU) 2413-357-00203447-99 in the amount of 10 weight fractions were considered as the research object. The modification was carried out when introducing мetal/C nanocomposite into the ER.

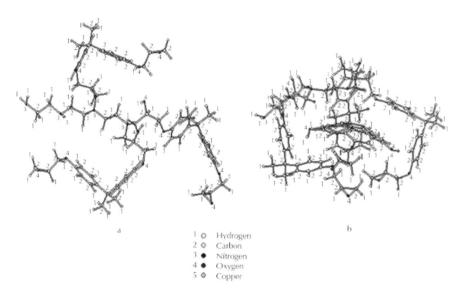

1	○	Hydrogen
2	◎	Carbon
3	●	Nitrogen
4	●	Oxygen
5	◎	Copper

FIGURE 4   Influence of Cu/C NC on ER hardening: (a) ER hardening without Cu/C NC; (b) ER hardening in the presence of Cu/C NC.

The PEPA represents the mixture of linear branched ethylene polyamines with average molecular mass 200–250 and very wide molecular-mass distribution. The general structural formula of PEPA is as follows:

$$NH_2[-CH_2CH_2NH-]_n-H$$

where $n = 2-8$.

Synthesized Co/C, Ni/C, and Cu/C nanocomposites were studied with the help of transmission electron microscopy (TEM) and electron microdiffraction (EMD) in the shared centers of research institutions in Ekaterinburg and Moscow.

The investigations of Cu/C and Ni/C nanocomposites revealed that their average sizes (r) differ approximately in two times: r of Cu/C NC is 25 nm, r of Ni/C NC is about 11 nm. The average size of Co/C nanocomposite practically equals the average size of Ni/C NC. Such correlation of average sizes of nanocomposites is apparently connected with the ability to form metal containing clusters. Atom magnetic moments of nanocomposites are greater than the corresponding moments of microparticles of the same metals [2]. The availability of magnetic properties of nanoproducts widens their application possibilities.

5.2.2   Processing of Fine Suspensions
The FS was prepared in a number of stages:
1.   Preliminary grinding of Cu/C NC in mechanical mortar.
2.   Mechanical and chemical activation of suspension components at the time of combining PEPA and Cu/C nanocomposite.

3. Ultrasound processing for complete and uniform dispergation of Cu/C NC particles in PEPA volume.

5.2.3 Processing of Epoxy Resins Modified with Cu/C Nanocomposite

The samples of epoxy polymer modified with Cu/C NC were produced mixing the ER heated up to 60°C and FS of Cu/C NC on PEPA basis in proportion 10:1. The mixing took 5–10 min. A part of the polymer modified with NC was poured into the mold with copper wire to test the adhesive strength (Figure 5) and was further hardened in the mold. Another part was hardened in the form of plates to grind the samples for thermo gravimetric investigation.

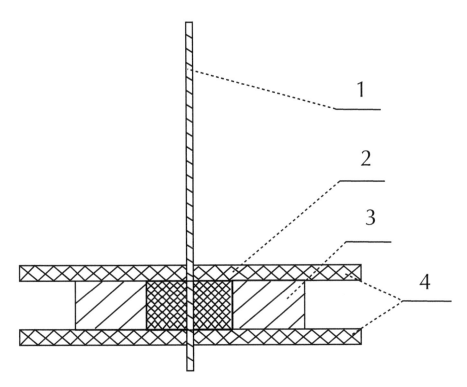

FIGURE 5 Sample for determining the adhesive strength of cold-hardened epoxy composition (CHEC): (1) copper wire; (2) hardened composition; (3) metal mold; and (4) anti-adhesive liner.

Two sets of samples were produced to determine the adhesive strength, including the reference and modified FS of Cu/C NC with concentrations 0.001, 0.01, and 0.03%. For oxide film removal and degreasing, the copper wire was treated with 0.1 M solution of hydrochloric acid and acetone.

Also for thermogravimetric (TG) investigation, two sets of samples were produced, including the reference and modified FS of Cu/C NC with concentrations 0.001, 0.01, 0.03, and 0.05%.

5.2.4   Investigation Technique

1. *Quantum-chemical Modeling*: In the frameworks of this method, the fragments imitating the behavior of Co/C, Ni/C, and Cu/C nanocomposites, FS of nanocomposites and systems being ERs modified with nanocomposites were constructed and optimized with the help of software HyperChem v. 6.03 and semi-empirical methods. At the same time, the absolute binding energies of certain components of the molecular systems formed were found and their relative interaction energies were calculated. The oscillatory spectra of FS of Cu/C nanocomposite were calculated with the help of semi-empirical method PM3.

2. *IR Spectroscopy*: For IR spectra IR Fourier-spectrometer FSM 1201 was used. The IR spectra of liquid films of PEPA, NC FS with concentrations 0.001, 0.01, and 0.03% were taken. The spectra were taken on KBr glasses in wave number range 399–4,500 cm^{-1}.

3. *Optical Spectroscopy*: Spectral photometer KFK-03-01 was used to define the optimal time period of NC FS ultrasound processing. The optimal time period value was the interval when the optical density (D) was at the maximum that is corresponded to the maximum saturation of FS with NC. The optical density of NC FS samples with the concentration 0.01% processed in ultrasound bath Sapfir UZV 28 within 0, 3, 7, 10, 15, 20, and 30 min, respectively, at US power 0.5 Kw and frequency 35 kHz was measured in 5 ml quartz cuvettes. The work wavelength was found when defining the optical density in the range $\lambda = 320–920$ nm.

4. *Determination of Relative Viscosity*: The relative viscosity of PEPA and FS on its basis was found with the help of viscometer VZ-246 in accordance with State Standard (GOST) 8420-74. The relative viscosity values were translated into the kinematic following GOST 8420-74. The viscosity was measured for PEPA processed and not processed with ultrasound, and also for FS of NC with the concentrations 0.001, 0.01, and 0.03% processed and not processed with ultrasound.

5. *Thermogravimetric Technique*: Thermal stability was found by the destruction temperatures of modified ERs. The temperatures of destruction beginning were determined by TG curves. Derivatographer DIAMOND TG/DTA was applied to obtain TG curves. The TG curves of modified ERs with NC concentrations 0.001, 0.03, and 0.05% from PEPA weight were taken. The sample heating rate was 5°C/min.

6. *Determination of Adhesive Strength*: The adhesive strength of NC/EC was found with the technique described (Figure 5). The tests were carried out on the tensile testing machine. The strength was found comparing breaking stresses of CHEC and NC/EC samples.

5.3   DISCUSSION AND RESULTS

5.3.1   Optimal Time Period of Ultrasound Processing of Fine Suspensions of Copper/Carbon Nanocomposite

In the process of selecting the work wavelength (λ) on spectral photometer KFK-03-01 to define the optimal time period of ultrasound processing of FS on PEPA basis, the optimal wavelength 413 nm was found. This wavelength was selected as the optical density of FS with the concentration of Cu/C NC 0.01% not processed with ultrasound was 0.800 that corresponded to the middle of the operational range of this instrument (operational range D = 0.001–1.5). At the wavelength λ = 413 nm, the optical densities of all FS investigated were defined. This is reflected in the graphic dependence of optical density D upon the time period of ultrasound processing τ_{us} (Figure 6).

The analysis of optical density dependence upon the time period of ultrasound processing (Figure 6) demonstrates that the optimal processing time of Cu/C NC FS on PEPA basis is 20 min. Further ultrasound processing is useless as in 20 min the maximum optical density 1.37 is reached.

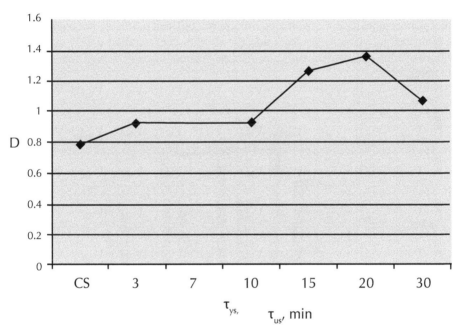

FIGURE 6 Dependence of the change in optical density (D) of Cu/C NC FS upon the time period of ultrasound processing τus.

Taking into account the above data for defining the optimal time period of ultrasound processing, the preparation of FS for IR investigations, for finding the viscosity and for producing the modified ER was carried out with preliminary ultrasound processing within 20 min.

5.3.2 Viscous Properties of Fine Suspensions of Copper/Carbon Nanocomposite

The following diagram was prepared based on the results of viscosity measurements of PEPA and FS with different Cu/C NC content before and after US processing (Figure 7).

The decrease in FS viscosity with the increase in Cu/C NC concentration on the diagram (Figure 7) is explained by the fact that the system consisting of two phases, in our case (PEPA—dispersion medium and Cu/C NC—disperse phase) tends to surface energy decrease. This tendency is expressed by self-decrease in interface surface due to sorption. The PEPA molecules start sorbing on Cu/C NC surface, probably producing the interface between Cu/C NC particles being the obstacle for coagulation. Due to such localization, the interaction between medium particles diminishes, resulting in intermolecular friction decrease, and, consequently, decrease in system viscosity (Figure 8(a)). When FS are processed with ultrasound, the size of disperse phase decreases, thus leading to the growth of surface energy and increase in sorption ability, the action region on disperses medium increases (Figure 8(b)). The action regions of Cu/C nanocomposite in FS with concentrations 0.001 and 0.01% do not overlap that is confirmed by viscosity decrease and for FS with concentration 0.03% the action regions overlap and, consequently, the increase in intermolecular friction and viscosity are observed.

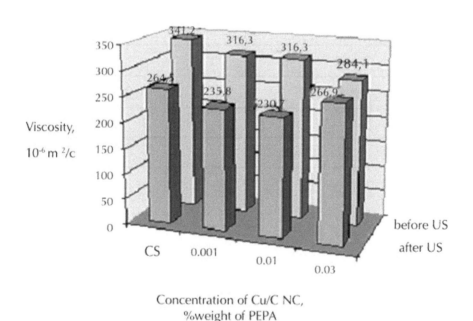

FIGURE 7   Diagram of the dependence of FS kinematic viscosity on Cu/C NC concentration.

Taking into account that the action degree on the medium mainly depends on the action time of Cu/C NC on the medium modified, FS were studied for 48 hr. After this time interval, FS layered into floccular structures with no sediment on the bottom observed (Figure 8(c)).

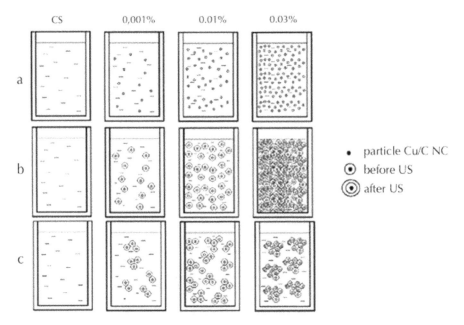

FIGURE 8   Distribution of Cu/C NC in PEPA: (a) before ultrasound processing; (b) after ultrasound processing; (c) after 48 hr.

Mechanical action resulted in FS recovery, which is indirectly confirmed by the availability of PEPA stable layer between Cu/C nanocomposite particles preventing the coagulation and possibility of suspension production with the ability to recover the distribution of nanocomposite particle distribution.

5.3.3   IR Spectroscopic Investigation

The comparative IR spectroscopic investigation of bis(2-aminoethyl)amine one of PEPA olygomers demonstrates the consistency of wave numbers appropriate for these compounds.

In both spectra, the wave numbers appropriate for the oscillations of amine groups $v_s(NH_2)$ 3,352 cm^{-1} and asymmetric $v_{as}(NH_2)$ 3,280 cm^{-1} are available, there are wave numbers that refer to symmetric $v_s(CH_2)$ 2,933 cm^{-1} and asymmetric valence $v_{as}(CH_2)$ 2,803 cm^{-1}, deformation wagging oscillations $v_d(CH_2)$ 1,349 cm^{-1} of methylene groups, deformation oscillations v_d (NH) 1,596 cm^{-1} and $v_d(NH_2)$ 1,456 cm^{-1} of amine groups, and also the oscillations of skeleton bonds are vivid $v(CN)$ 1,059–1,273 cm^{-1} and $v(CC)$ 837 cm^{-1}.

The IR spectra of PEPA and Cu/C NC FS were taken (Figure 9).

The comparison of IR spectra of PEPA and FS of Cu/C NC on PEPA basis (Figure 9) indicates that practically all changes of wave numbers in the spectra are within the error ± 2 cm^{-1}. However, in FS spectra the vivid increase in peak intensity corresponding to deformation oscillations of NH bonds is observed. These changes can spread onto the vast areas arranging a certain supermolecular structure, apparently involving the adjoining amine groups into the process, which is demonstrated by the intensity change of these peaks.

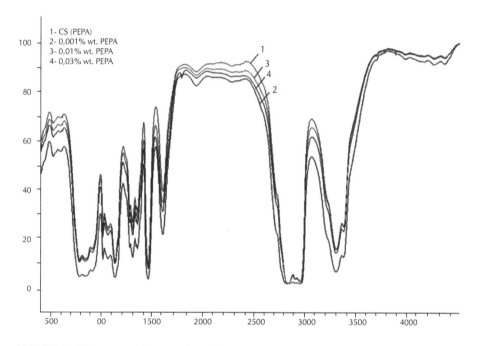

FIGURE 9   IR spectra of PEPA and FS of Cu/C nanocomposite.

5.3.4   Adhesion

The tests for defining the adhesion of modified ER to copper wire were carried out on tensile testing machine, the values of destruction load were found. The adhesive strength was calculated by the following formula:

$$\sigma = F/A \text{ [MPa]}$$

where F = average load values at which the breaking-off took place, [kgs]; A = area of wire interaction with the hardened composition, [cm²]

$$A = 2\pi rh = 2 \times 3.14 \times 0.05 \times 1 = 0.314 \text{ [cm}^2\text{]}$$

where r = wire radius 0.05 [cm]; h = height of metal mold 1.00 [cm].
    The following diagram was prepared based on tests (Figure 10)

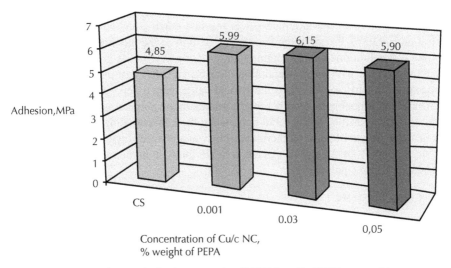

FIGURE 10 Dependence of adhesive strength of NC/EC on Cu/C NC composition.

5.3.5 Thermogravimetric Investigations

To define the influence of nanocomposite on thermal stability of epoxy composition, a number of TG investigations were carried out on reference and modified samples. The concentrations 0.001, 0.03, and 0.05% from PEPA weight were used. Based on the results of TG investigations, the following diagram was prepared (Figure 11).

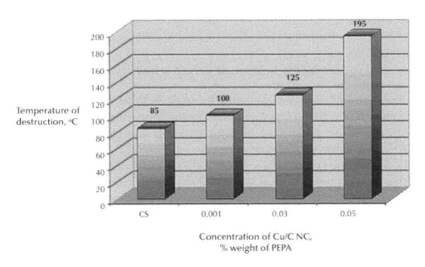

FIGURE 11 Dependence of thermal stability of CHEC on the concentration of copper/carbon nanocomposite.

The data from the diagram of adhesive strength (Figure 10) indicate that the strength maximum of the modified ER is reached when Cu/C nanocomposite concentration is 0.03%. Probably the maximum strength of the modified ER at this concentration is conditioned by the optimal number of a new phase growth centers. The adhesion decrease with the concentration increase indicates that the number of Cu/C nanocomposite particles exceeded the critical value, which depends on their activity [2]. Therefore, probably the number of cross-links in the polymer grid increased—the material became brittle. At the same time, when the concentration of nanocomposite elevates, the growth of polymer thermal stability is observed (Figure 11). The thermal stability growth is apparently connected with the increase in the number of coordination bonds in epoxy polymer.

5.4   CONCLUSION

Quantum-chemical modeling methods allow quite precisely defining the typical reaction of interaction between the system components, predict the properties of molecular systems, and decrease the number of experiments due to the imitation of technological processes. The computational results are completely comparable with the experimental modeling results.

The optimal composition of nanocomposite was found with the help of software HyperChem v. 6.03. Cu/C nanocomposite and is most effective for modifying the ER. Nanosystems formed with this NC have higher interaction energy in comparison with nanosystems produced with Ni/C and Co/C nanocomposites. The effective charges and geometries of nanosystems were found with semi-empirical methods. The fact of producing stable FS of nanocomposite on PEPA basis and increasing the operational characteristics of epoxy polymer was ascertained. It was demonstrated that the introduction of Cu/C nanocomposite into PEPA facilitates the formation of coordination bonds with nitrogen of amine groups, thus resulting in PEPA activity increase in ER hardening reactions.

It was found out that the optimal time period of ultrasound processing of copper/carbon nanocomposite FS is 20 min.

The dependence of Cu/C nanocomposite influence on PEPA viscosity in the concentration range 0.001–0.03% was found. The growth of specific surface of NC particles contributes to partial decrease in PEPA kinematic viscosity at concentrations 0.001, 0.01% with its further elevation at the concentration 0.03%.

The IR investigation of Cu/C nanocomposite FS confirms the quantum-chemical computational experiment regarding the availability of NC interactions with PEPA amine groups. The intensity of these groups increased several times when Cu/C nanocomposite was introduced.

The test for defining the adhesive strength and thermal stability correlate with the data of quantum-chemical calculations and indicate the formation of a new phase, facilitating the growth of cross-links number in polymer grid when the concentration of Cu/C nanocomposite goes up. The optimal concentration for elevating the modified ER adhesion equals 0.003% from ER weight. At this concentration, the strength growth is 26.8%. From the concentration range studied, the concentration 0.05% from

ER weight is optimal to reach a high thermal stability. At this concentration, the temperature of thermal destruction beginning increases up to 195°C.

Thus, in this work the stable FS of Cu/C nanocomposite were obtained. The modified polymers with increased adhesive strength (by 26.8%) and thermal stability (by 110°C) were produced based on ERs and FS.

KEYWORDS

- **Fine suspensions**
- **Nanocomposite**
- **Nanostructures**
- **Polymeric composite materials**
- **Quantum-chemical modeling**
- **Thermogravimetric**

REFERENCES

1. Panfilov, B. F. Composite materials: Production, application, market tendencies. *Polymeric Materials,*, (2–3),40–43 (2010).
2. Kodolov, V. I., Khokhriakov, N. V., Trineeva, V. V., and Blagodatskikh I. I. Activity of nanostructures and its expression in nanoreactors of polymeric matrices and active media. *Chemical Physics and Mesoscopy*, **10**(4), 448–460 (2008).
3. Bobylev, V. A. and Ivanov, A. V. New epoxy systems for glues and sealers produced by CJSC "Chimex Ltd". *Glues. Sealers. Paints*, (2), 162–166 (2008).
4. Kodolov, V. I., Kovyazinam O. A., Trineeva, V. V., Vasilchenko, Yu. M., Vakhrushina, M. A., and Chmutin, I. A. On the production of metal/carbon nanocomposites, water and organic suspensions on their basis. VII International Scientific-Technical Conference "Nanotechnologies to the production—2010". Proceedings, Fryazino, 52–53 (2010).

6 Advances on Permeability of Polymer Nanocomposites

B. Zh. Dzhangurazov, G. V. Kozlov,
G. E. Zaikov, and A. K. Mikitaev

CONTENTS

6.1 INTRODUCTION

Over the last 15 years, the development of nanocomposites polymer/organoclay evokes a great interest owing to considerable improvement of their physical and mechanical properties in comparison with pristine matrix polymer at small (no more than 10 mass %) nanofiller contents [1, 2]. One of the indicated changes of these nanocomposites properties is the essential reduction of permeability to gas coefficient P. So, in number of works [3], it has been shown that the introduction of montmorillonite (MMT) into polyethylene (PE) at the volume content $\varphi_n = 0.065$ of the latter decreases the value P in several times in comparison with the pristine matrix polymer. This effect analysis, fulfilled by the authors [3], assumes that permeability to gas reduction is connected not with matrix PE structure change at MMT introduction but with meandering way of considerable increasing of gas molecules at diffusion through nanocomposite film, containing anisotropic MMT particles. It is significant that [4, 5] structural analysis of the indicated nanocomposites have also shown polymer matrix structure invariability at MMT introduction. For instance, the multifractal model of gas transport processes [6] allows the quantitative estimation of meanderibility degree of gas-penetrant molecules way through polymeric material. Therefore, the purpose of present chapter is the structural analysis of nanocomposites PE/MMT permeability to gas coefficient reduction effect with multifractal model [6] usage.

6.2 EXPERIMENTAL

The experimental values of permeability to gas coefficient P by three gas-penetrants $(N_2, O_2,$ and $CO_2)$, acquired in paper [3], were used for nanocomposites PE/MMT with MMT contents $\varphi_n = 0.008, 0.016,$ and 0.065. The value P was determined on a baric device [7] at pressure 1–2 MPa and temperature 293K. Nanocomposites samples of thickness 70–80 mcm were prepared by pressing at temperature 458K and pressure 30 MPa.

6.3 RESULTS

The basic equation of gas transport processes fractal model has the following form [8]:

$$P = P_0 f_g \left(d_h / d_m \right)^{2(D_t - d_s)/d_s}, \tag{1}$$

where P_0 is constant, f_g is relative fluctuation free volume, d_h and d_m are diameters of free volume microvoid and gas-penetrant molecule, respectively, D_t is structure dimension, controlling gas transport processes, d_s is spectral dimension, which is accepted equal to 1.0 for linear PE [9].

From the Equation (1), it follows that the dependence $P(1/d_m)$ at $d_h =$ const and $d_s =$ const plotting should give linear correlation, from the slope of which the exponent $2(D_t-d_s)/d_s$ value in the Equation (1) can be determined and, consequently, the dimension D_t. As it has been shown earlier [8], for polymeric materials two mechanisms of gas transport can be realized: structural or molecular one, depending on the ratio (d_h/d_m) value. At $(d_h/d_m) > 1.7$, the interaction of gas-penetrant molecules and free volume microvoids walls is absent and in this case the value D_t is accepted equal to fractal (Hausdorff) dimension of structure d_f. At $(d_h/d_m) \leq 1.7$ gas-penetrant molecule interacts with the indicated microvoids walls, whose dimension is approximately equal to the dimension of excess energy localization regions D_f and then $D_t = D_f$. It is obvious that for the first mechanism, the general for all fractal objects restriction exists: $d_f \leq d$, where d is the dimension of Euclidean space in which fractal is considered.

The plotting of the indicated dependence $P(1/d_m)$ for pristine PE and nanocomposites PE/MMT with MMT contents $\varphi_n = 0.008, 0.016,$ and 0.065 is adduced in Figure 1, from which it follows, that these dependences are actually linear, their slope for four studied materials is the same and value $D_t \approx 6.6$. From the above mentioned proceeding, one should accept $D_n = D_f$. As it is known [8], the dimensions D_f and d_f are connected between themselves by the following relationship:

$$D_f = 1 + \frac{1}{3 - d_f} \tag{2}$$

From the Equation (2), it follows that $d_f = 2.82 =$ constant. The indicated d_f value is close to the obtained ones by other methods for PE [8] and the condition $d_f =$ constant shows that MMT introduction does not change pristine matrix PE structure.

It has been shown earlier [10] that P value for semi-crystalline polymer can be written as follows:

$$P = \frac{P_{am}}{\tau\beta},$$ (3)

where P_{am} is permeability to gas coefficient of completely amorphous polymer, t is meanderibility coefficient, which is due to complexity of gases transport ways between crystallites and depending not only on crystallinity degree K but on value, shape, and distribution of crystallites (and nanofiller particles in the considered case) by sizes, b is chains immobility coefficient, which depends on molecular mobility level and assumed as constant in virtue of the condition d_f = constant.

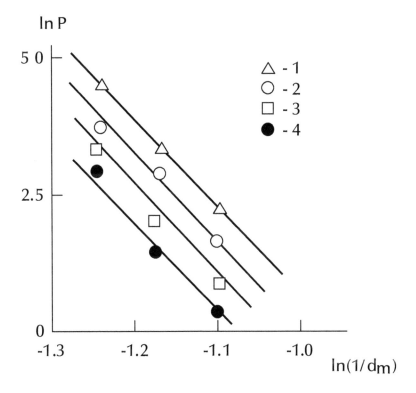

FIGURE 1   The dependence of permeability to gas coefficient P on reciprocal value of gas-penetrant molecule diameter d_m in double logarithmic coordinates for matrix PE (1) and nanocomposites PE/MMT with MMT contents φ_n = 0.008 (2), 0.016 (3), and 0.065 (4).

In paper [6], it has been shown that within the frameworks of gas transport processes multifractal model the value t is defined as follows:

$$P = \frac{P_{am}}{\tau\beta},$$ (4)

where α_{am}^{ac} is amorphous phase fraction, α_{am} is accessible for diffusion of gas with molecule diameter d_m, which is equal to as follows [6]:

$$\alpha_{am}^{ac} = \left(\alpha_{am}\right)^{d_m}$$ (5)

d_m values for the studied gases are adduced in paper [11] and the value α_{am} in case of the considered nanocomposites can be determined in the following way. The first of them assumes that crystallites and MMT particles are impenetrable for gases diffusion as in the following:

$$\alpha_{am} = 1 - K - \varphi_n,$$ (6)

where value K is accepted equal to 0.60 [3].

The second mode assumes that densely packed on layered exfoliated MMT surface interfacial regions with relative fraction φ_{if} are also impenetrable for diffusion and in this case, it is as follows:

$$\alpha_{am} = 1 - K - \varphi_n - \varphi_{if}$$ (7)

The relation between φ_{if} and φ_n for the exfoliated MMT has the following appearance [5]:

$$\varphi_{if} = 1.91\varphi_n$$ (8)

Then relative permeability to gas coefficient P_n/P_{PE} (where P_n and P_{PE} are permeability to gas coefficients of nanocomposites PE/MMT and matrix PE, respectively) can be written as follows:

$$\frac{P_n}{P_{PE}} = \left(\frac{\alpha_{am}^n}{\alpha_{am}^{PE}}\right)^{d_m},$$ (9)

where α_{am}^n and α_{am}^{PE} are amorphous phase relative fractions for nanocomposites PE/MMT and matrix PE, respectively.

In Figure 2, the comparison of experimental (points) and calculated according to the indicated mode (with the Equation (7) usage) values P_n/P_{PE} (curve 1) is adduced. As one can see, the well enough correspondence of theory and experiment was obtained. In the same Figure, the theoretical dependence of P_n/P_{PE} on φ_n is adduced, which does not take into consideration impenetrable nature of interfacial regions for gases diffusion, that is its calculation the Equation (6) was used (curve 2). In this case, for clear reasons, overstated theoretical values of permeability to gas coefficient for the studied nanocomposites are obtained.

One can improve theory (namely, curve 1) and experiment correspondence by the usage of the notion about nanoadhesion [5]—the conception, which is specific for nanocomposites. The Equation (8) was derived in assumption of perfect microadhesion and it can be written in more general form, taking into consideration interfacial adhesion level on polymer matrix—nanofiller boundary, as follows:

$$\varphi_{if} = 1.91\varphi_n b,$$ (10)

where b is parameter, accounting for interfacial adhesion level. At $b = 1.0$, the perfect microadhesion is observed, at $b > 1.0$ nanoadhesion. As it is known [4, 5], nanocomposites reinforcement degree E_n/E_{PE} can be determined according to the following equation:

$$\frac{E_n}{E_{PE}} = 1 + 11\left(\varphi_n + \varphi_{if}\right)^{1.7},$$ (11)

where E_n and E_{PE} are elasticity moduli of nanocomposite and matrix PE, respectively.

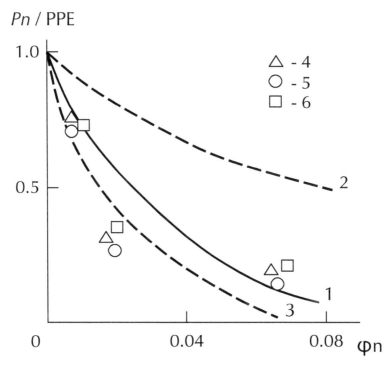

FIGURE 2   The dependence of relative permeability to gas coefficient P_n/P_{PE} on MMT volume contents φ_n for nanocomposites PE/MMT. Calculation: (1) according to the Equations (9) and (7); (2) according to the Equations (9) and (6); (3) according to the Equations (9) and (10). The experimental data for N_2 (4), O_2 (5), and CO_2 (6).

According to the data from paper [3], $E_n = 680$ MPa and $E_{PE} = 510$ MPa. Then the Equation (11) allows obtaining the following relationship for the studied nanocomposites:

$$\varphi_{if} = 3.27\varphi_n \qquad (12)$$

Let us obtain $b = 1.71$ from the comparison of the Equations (8), (10), and (11). Therefore, for the studied nanocomposites the nanoadhesion effect is realized, although a weak enough one. Let us remind that for nanocomposites phenylone/aerosil, the value b can reach 13 [5]. Nevertheless, considering even such a weak nanoadhesion effect, that is the Equation (10) usage allows to improving essentially theory and experiment correspondence (curve 3) and their discrepancy at $\varphi_n = 0.065$ is explained by MMT incomplete exfoliation at its high enough contents [5].

6.4 CONCLUSION

Therefore, the proposed structural model describes well nanocomposites PE/MMT permeability to gas reduction at nanofiller contents increasing only at taking into consideration interfacial regions role in this effect. Taking into consideration the nanoadhesion effect, connected with interfacial regions relative fraction, allows improving the indicated correspondence.

KEYWORDS

- **Gas-penetrant molecule**
- **Montmorillonite**
- **Nanoadhesion**
- **Nanocomposites polymer**
- **Polyethylene**

REFERENCES

1. Kojima, Y., Usuki, A., Kawasumi, M., Okada, A., Kurauchi, T., and Kamigaito, O. *J. Polymer Sci.*: Part A: *Polymer Chem.*, **31**(7), 1755–1769 (1993).
2. Kojima, Y., Usuki, A., Kawasumi, M., Okada, A., Kurauchi, T., and Kamigaito, O. *J. Appl. Polymer Sci.*, **49**(7), 1259–1265 (1993).
3. Kovaleva, N. Yu., Brevnov, P. N., Grinev, V. G., Kuznetsov, S. P., Pozdnyakova, I. V., Chvalun, S. N., Sinevich, E. A., and Novokshonova, L. *A. Vysokomolek. Soed. A*, **46**(6), 1045–1051 (2004).
4. Kozlov, G. V., Malamatov, A. Kh., Antipov, E. M., Karnet, Yu. N., and Yanovskii, Yu. G. *Mekhanika Kompozitsionnykh Materialovi Konstruktsii*, **12**(1), 99–140 (2006).
5. Mikitaev, A. K., Kozlov, G. V., and Zaikov, G. E. *Polymer Nanocomposites: The variety of structural forms and applications*. Nova Science Publishers, Inc., New York, p. 319 (2008).
6. Khalikov, R. M. and Kozlov, G. V. *Vysokomolek. Soed. B*, **48**(4), 699–703 (2006).
7. Sinevich, E. A., Arzhakov, M. S., Bykova, I. S., Krykin, M. A., Shitov, N. A., Timashov, S. F., and Bakeev, N. F. *Vysokomolek. Soed. A*, **36**(1), 123–127 (1994).

8. Kozlov, G. V., Zaikov, G. E., and Mikitaev, A. K. *The fractal analysis of gas transport in polymers: The theory and practical applications.* Nova Science Publishers, Inc., New York, p. 238 (2009).
9. Alexander, S. and Orbach, R. *J. Phys. Lett. (Paris)*, **43**(17), 625–L631 (1982).
10. Rogers, C. E. In *Engineering Design for Plastics.* E. Baer (Ed.). Reinhold Publishing Corporation Chapman and Hall, LTD, London, pp. 193–273 (1966).
11. Teplyakov, V. V. and Durgar'yan, S. G. *Vysokomolek. Soed. A*, **24**(7), 1498–1505 (1984).

7 A Note on Hybrid Nanocomposites

I. Yu. Yevchuk, O. I. Demchyna, Yu. G. Medvedevskikh, G. E. Zaikov, H. V. Romaniuk, and I. I. Tsiupko

CONTENTS

7.1 INTRODUCTION

The creation of nanocomposite materials on the base of hybrid organic-inorganic systems is one of the perspective directions of polymeric material engineering. Nanocomposite materials are increasingly applied in technique, medicine, biotechnology, and so on because synthesized materials often have unique properties due to inorganic particles with polymeric matrix on nanosize level [1, 2].

Among many methods of obtaining such materials, sol-gel technology deserves special attention. The advantages of this method are as followshomogeneous dispersion of inorganic component in polymeric matrix, energy saving regime (processes take place at low temperatures), ecological safety (in sol-gel processes mainly water-alcohol solutions of precursors are used), possibility of purposeful introduction of modifiers during the stage of synthesis, and conducting of template synthesis. On using sol-gel method materials of various applications may be obtainedthermo sensitive, magnetic, nonlinear-optical, electro chromic, insulating composites, ion-conducting materials, gas-diffusion electrodes, membranes, nanocatalysts, bioactive compounds, and so on. [3, 4].

Initial components for synthesis of materials by sol-gel method are Si, Ti, Al, B, W, Zr, and Mo alkoxides and others with functionality of 3 or 4, which hydrolyze at addition of water. Following polycondensation leads to gel formation. If sol-gel transformation takes place in organic polymer medium, inorganic structure is formed *in situ* in polymeric matrix. Obtaining organic-inorganic materials is also possible during simultaneous conducting of the processes of organic monomers polymerization and forming of nanosize particles of inorganic phase as a result of sol-gel process.

The analysis of numerous publications on obtaining organic-inorganic composites indicates that the main tendency in this research field is accumulation of experimental information and establishment of dependence between the way of synthesis, structure, and properties of material. However, the problem of obtaining materials with the adjusted characteristics by the guided synthesis is far from its solution. Therefore, actual research task is investigation of different factors influence on the processes of syntheses of hybrid materials with the purpose to obtain products with desirable properties.

7.2   EXPERIMENTAL

For our investigations, we have used the following reagents tetraethoxysilane $Si(OC_2H_5)_4$ ("EKOS-1", Russia, technical specification 2637-059-444493179-04), ethanol ("r.g."), orthophosphoric acid ("r.g."), distilled water, and polyvinilidene fluoride (PVDF) M_w 175000 (Aldrich).

The rheological measurements were conducted by means of rotary viscosimeter "RHEOTEST-2.1" (VEB MLW, DDR). The relation between a shift tension and a shift rate was measured for sol-gel systems. The latter was regulated by the rate of rotation of measuring cylinder or cone. For support of stationary temperature, the working cylinder/cone with investigated system was placed into thermostatic volume. Dynamic viscosity was determined after correlation as follows:

$$\eta = \tau_r / D_r, \tag{1}$$

where η is dynamic viscosity ($Pa \cdot s$); τ_r is shift tension (Pa); D_r is shift rate (s^{-1}). Dependence of the measured viscosity on angular rotation rate ω of working cylinder is described by the following expression:

$$\eta = \eta_o + \eta_s \frac{1 - \exp\{-b / \omega\}}{1 + \exp\{-b / \omega\}} \tag{2}$$

where η_o is friction constituent of viscosity; η_s is elastic constituent of viscosity, and b is coefficient, which characterizes gradient dependence of viscosity. Parameters η_o, η_s, and b were determined by optimization method in origin 5.0 program.

Proton conducivity of the samples was measured by impedance spectrometer "autolab" (Ecochem, Holland) with FRA program. The investigated samples were gripped between two platinum electrodes of diameter of 1 cm. The thickness of the samples is ~1 mm. Impedance hodographs in the range of frequencies $10–10^5$ Hz were analyzed. As the value of proton conductivity, a value $1/R_F$ was taken where R_F is a cut-off on the axis of real resistance [5]. Specific proton conductivity was determined as reciprocal to specific resistance that was calculated after a formula as follows:

$$\rho = R \pi d^2 / 4l \tag{3}$$

where R is material resistance, Ohm; l is sample thickness, cm; and d is thickness of sample.

7.3   RESULTS

As it was mentioned above, for possibility to conduct sol-gel synthesis with the purpose of obtaining products with definite properties, it is important to know the kinetic regularities of processes that take place there. It should be noted that the problem of sol-gel syntheses kinetics is not yet investigated sufficiently. Silicate gels are synthesized usually by hydrolysis of monomer tetra functional alkoxide precursors using mineral acid (for example, HCl) or base (for example, NH_3) as catalyst. As a rule, three reactions are used for sol-gel process description on the functional groups levelhydrolysis ↔ esterification (Equation (4)), water condensation ↔ hydrolysis (Equation (5)), and alcohol condensation ↔ alcoholysis (Equation (6)) as in the following:

$$(RO)_3Si–OR + HOH \leftrightarrow (RO)_3Si–OH + ROH \qquad (4)$$
$$\equiv Si–OH + HO–Si\equiv \leftrightarrow \equiv Si–O–Si\equiv + HOH \qquad (5)$$
$$\equiv Si–OH + RO–Si\equiv \leftrightarrow \equiv Si–O–Si\equiv + ROH \qquad (6)$$

In reaction of hydrolysis (Equation (4)), alkoxide groups –OR are replaced with hydroxyl groups –OH. Subsequent condensation reactions involving silanol groups produce siloxane bonds (Si–O–Si) and by products – alcohol (ROH) (Equation (6)) or water (Equation (5)). Nanosize products of condensation of organosilanes of the general formula $SiO_x(OH)_y(OR)_z$ form silica framework uniting between themselves. Structuring of sol-gel system begins in so called percolation point, when such concentration and such conformational volumes of macromolecules occur, that conformational volumes of macromolecules begin to overlap. It leads to sharp increasing of sol-gel system viscosity. Hence, studying dynamics of viscosity of sol-gel systems one can estimate characteristic times of achievement of percolation point and their dependence on such parameters as initial sol-gel system composition, catalyst concentration, and temperature.

It is known that solutions of macromolecules are not Newtonian liquids. Thus, their viscosity depends not only on friction forces between the layers of mobile liquid but also on shift deformation of macromolecules. It causes dependence of viscosity on the value of rate gradient of hydrodynamic flow, which is determined by angular rotation rate of working cylinder.

Viscosity of investigated sol-gel systems $TEOS:H_2O:C_2H_5OH$ was measured at different temperatures, concentrations of catalyst (Orthophosphoric acid) and initial compositions of systems within the range of rotation rates of working cylinder 0.5–243 ppm. The changes of parameters η_0, η_s, and b in time for investigated systems are presented in Figures 1–3.

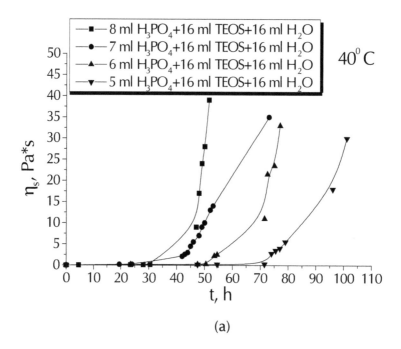

(a)

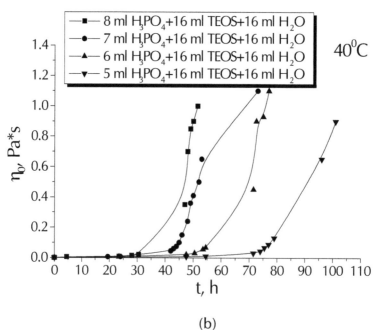

(b)

FIGURE 1 Dependence of η_s (a) and η_0 (b) of TEOS:C_2H_5OH:H_3PO_4:H_2O system on time at different concentrations of H_3PO_4 (40°C).

TABLE 1 Dependence of coefficient b of TEOS:C_2H_5OH:H_3PO_4:H_2O system on time at different H_3PO_4 concentrations (40°C).

8 ml H_3PO_4		7 ml H_3PO_4		6 ml H_3PO_4		5 ml H_3PO_4	
t, h	$b\times10^4$, s^{-1}	t, h	$b\times10^4$, s^{-1}	t, h	$b\times10^4$, s^{-1}	t, h	$b\times10^4$, s^{-1}
20	15.50	5	18.70	24	16.50	48	16.60
24	15.00	24	15.80	48	14.50	55	15.80
42	3.73	28	15.70	51	14.20	72	19.60
43	3.78	31	14.90	54	3.39	74	3.26
44	4.07	47	4.14	55	3.51	76	3.27
48	4.27	48	4.13	72	4.11	77	3.31
49	4.01	49	4.01	73	3.90	79	3.46
50	3.76	50	3.99	75	3.73	96	4.09
52	3.39	52	3.86	77	3.71	101	4.02

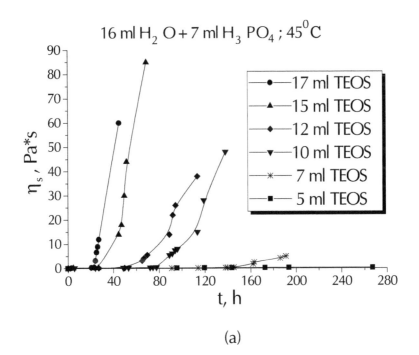

(a)

FIGURE 2 *(Continued)*

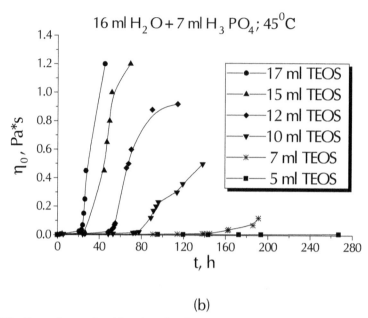

(b)

FIGURE 2 Dependence of ηs (a) and ηo (b) of TEOS:C_2H_5OH:H_3PO_4:H_2O system on time at different concentration of TEOS (45°C).

TABLE 2 Dependence of coefficient b of TEOS:C_2H_5OH:H_3PO_4:H_2O system on time at different concentrations of TEOS (45°C).

17 ml TEOS		15 ml TEOS		12 ml TEOS		10 ml TEOS		7 ml TEOS		5 ml TEOS	
t, h	$b\times 10^4, s^{-1}$	t, h	$b\times 10^4, s^{-1}$	t, h	$b\times 10^4, s^{-1}$	t, h	$b\times 10^4, s^{-1}$	t, h	$b\times 10^4, s^{-1}$	t, h	$b\times 10^4, s^{-1}$
3	17.80	5	18.80	26	16.80	6	18.00	91	19.60	95	17.80
20	18.80	22	17.10	48	17.70	24	16.50	114	17.80	119	18.00
23	20.00	26	16.50	50	18.10	48	16.20	138	17.20	143	18.40
24	3.28	44	3.30	51	3.13	53	16.10	139	17.60	172	17.40
25	3.07	47	3.15	52	3.17	72	16.80	145	17.30	193	17.10
26	3.00	49	3.12	53	3.26	74	16.90	162	3.61	267	16.50
27	3.03	51	3.14	54	3.28	77	17.30	163	3.31	—	—
45	4.42	69	3.73	55	3.51	89	2.83	186	2.89	—	—
—	—	—	—	65	2.93	91	2.84	191	2.76	—	—
—	—	—	—	67	2.96	94	2.92	—	—	—	—

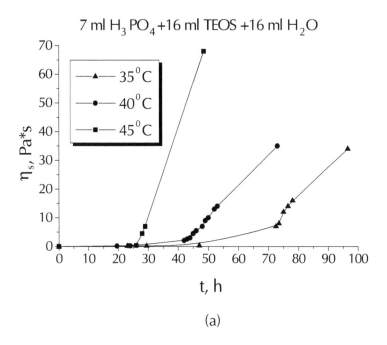

(a)

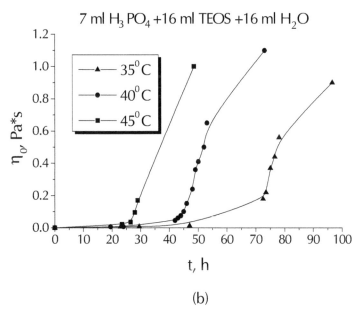

(b)

FIGURE 3 Dependence of η_s (a) and η_0 (b) of TEOS:C_2H_5OH:H_3PO_4:H_2O system on time at different temperatures.

TABLE 3   Dependence of coefficient b of system $TEOS:C_2H_5OH:H_3PO_4:H_2O$ on time at different temperatures.

35°C		40°C		45°C	
t, h	$B \times 10^4$, s^{-1}	t, h	$b \times 10^4$, s^{-1}	t, h	$b \times 10^4$, s^{-1}
23	17.10	20	15.50	24	16.00
30	16.90	24	15.00	27	14,10
47	14.10	42	3.73	28	2.61
73	3.85	43	3.78	29	2.74
74	3.89	44	4.07	49	3.59
75	3.08	48	4.27	—	—
77	2.98	49	4.01	—	—
78	2.94	50	3.76	—	—
97	2.90	52	3.39	—	—
—	—	53	3.28	—	—
—	—	73	2.49	—	—

As one can see in Figures 1–3, all curves consist of two parts. Slightly sloping part of the curve corresponds to gradual increasing of viscosity due to reactions of polycondensation of products of TEOS hydrolysis. Sharp changes of the slopes of the curves η_s—t and η_s—t are related with the beginning of gelation. Time of reaching of percolation point by the system was found after point of intersection of the tangents to these parts of the curve. In percolation point, conformational volumes of macromolecules begin to overlap, therefore sol-gel system becomes structured. It was found that time of percolation point reaching diminishes with the increase of TEOS and orthophosphoric acid concentrations in system as well as with the increase of temperature (Figure 4). The value of parameter b diminishes sharply. It testifies overlapping of conformational volumes of macromolecules as a result of gelation (Tables 1–3).

Hybrid organic-inorganic composites were synthesized by formation of nanoscale phase as a result of sol-gel process in systems on the basis of TEOS *in situ* in polymeric matrix of PVDF. This polymer was chosen due to its good operating characteristics high thermal and chemical stability. The PVDF was previously dissolved in *N,N'*-dimethylformamid (10 % mass.). Sol-gel system (sample 1: $TEOS:C_2H_5OH:H_3PO_4:H_2O$ = 2,2:7,34:0,1:0,36 v. p., sample 2: $TEOS:C_2H_5OH:H_3PO_4:H_2O$ = 2,2:7,24:0,2:0,36 v. p.) was added to PVDF solution to obtain the mixtures with PVDF/TEOS 70/30 (mass). Obtained mixtures were stirred for 2 hr at 40°C and exposed for film formation.

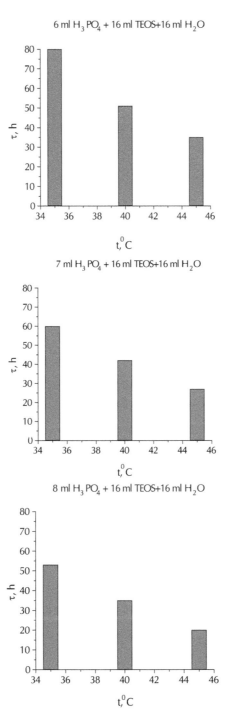

FIGURE 4   Times of achievement of percolation points in TEOS:C_2H_5OH:H_3PO_4:H_2O systems at different temperatures depending on H_3PO_4 concentration.

Proton conductivity of polymer nanocomposite films may be determined by measuring complex resistance–impedance; the active (real) and reactive (imaginary) constituents of vector of impedance allow to determine conducting characteristics of material. In Figure 5, hodographs of impedance of Pt sample—Pt cell for samples with different concentration of H_3PO_4

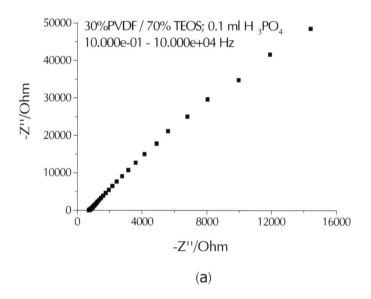

(a)

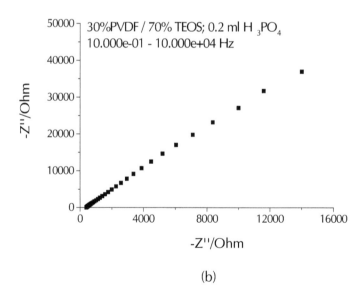

(b)

FIGURE 5   Hodographs of impedance for samples 1 (a) and 2 (b).

The measured proton conductivity of the samples is ~10^{-4} Sm/cm and depends on concentration of orthoposphoric acid in sample. Obviously, it is provided by silanol groups and P-OH groups.

7.4 CONCLUSION

The conducted research allows offering the synthesis of hybrid organic-inorganic materials by introducing sol-gel system on the basis of TEOS into the matrix of PVDF. Obtained nanocomposites have proton conductivity. Our further investigations will be conducted with the goal to provide sufficient increase of proton conductivity of nanocomposites for obtaining materials for proton-conducting membranes of fuel cells and gas sensors.

KEYWORDS

- **Nanocomposite**
- **Nanoscale**
- **Organic-inorganic materials**
- **Proton-conducting**
- **Sol-gel technology**

REFERENCES

1. Buchachenko, A. L. Nanochemistry as a direct way to high technologies of a new age. *Uspekhi khimiyi*, **72**(5), 419–436 (2003).
2. Pomogaylo, A. D. Hybrid polymer-inorganic nanocomposites. *Uspekhi khimii*, **69**(1), 60–89 (2000).
3. Shylova, E. A. and Shylov, A. A. Nanocomposite oxide and hybrid organo-inorganic materials, obtained by sol-gel method. Synthesis. Properties. Application. *Nanosystems, nanomaterials, nanotechnologies*, **1**(1), p. 9–83 (2003).
4. Roldugin, V. I. Self-assembling of nanopartiles on interphase surfaces. *Uspekhi khmiyi*, **73**(2), 123–156 (2004).
5. Dobrovolsky, Yu. A., Pysareva, A. V., Lenova, L. S., and Karelin, A. I. New proton-conducting membranes for fuel cells and gas sensors. *Alternative energetics and ecology*, **12**(20), 36–41 (2004).

8   Transformation of High-energy Bonds in ATP

G. A. Korable, N. V. Khokhriakov, G. E. Zaikov,
and Yu. G. Vasiliev

CONTENTS

8.1   INTRODUCTION

With the help of spatial-energy concept, it is demonstrated that the formation and change of high-energy bonds in ATP takes place at the functional transitions of valence active orbitals of the "phosphorus-oxygen" system.

These values of energy bonds are in accordance with experimental and quantum mechanical data.

8.2   SPATIAL-ENERGY PARAMETER

During the interaction of oppositely charged heterogeneous systems, certain compensation of volume energy of interacting structures take place, which leads to the decrease in the resulting energy (e.g., during the hybridization of atomic orbital). But this is not the direct algebraic deduction of the corresponding energies. The comparison of multiple regularities of physical, chemical, and biological processes allow assuming that in such and similar cases the principle of adding the reciprocals of volume energies or kinetic parameters of interacting structures is executed.

Lagrangian equation for the relative movement of the system of two interacting material points with the masses m_1 and m_2 in coordinate x is as follows:

$$M_{red}x'' = -\frac{\partial U}{\partial x}, \text{ where } \frac{1}{m_{red}} = \frac{1}{m_1} + \frac{1}{m_2} \qquad \text{(1) and (1a)}$$

Here, U = mutual potential energy of material points; m_{red} = reduced mass. Herein, $x'' = a$ (system acceleration).

For the elementary areas of interactions Δx, we can accept: $\dfrac{\partial U}{\partial x} \approx \dfrac{\Delta U}{\Delta x}$

Then

$$m_{red} a \Delta x = -\Delta U; \qquad \frac{1}{1/(a\Delta x)} \cdot \frac{1}{(1/m_1 + 1/m_2)} \approx -\Delta U$$

or

$$\frac{1}{1/(m_1 a\Delta x) + 1/(m_2 a\Delta x)} \approx -\Delta U$$

Since the product $m_i a \Delta x$ by its physical sense equals the potential energy of each material point (ΔU_i), then

$$\frac{1}{\Delta U} \approx \frac{1}{\Delta U_1} + \frac{1}{\Delta U_2} \qquad (2)$$

Thus, the resulting energy characteristic of the system of two interacting material points is found by the principle of adding the reciprocals of initial energies of interacting subsystems.

"The electron with the mass m moving about the proton with the mass M is equivalent to the particle with the mass: $m_{red} = \dfrac{mM}{m+M}$ [1]"

Therefore, modifying the Equation (2), we can assume that the energy of atom valence orbitals (responsible for interatomic interactions) can be calculated [2] by the principle of adding the reciprocals of some initial energy components based on the following equations:

$$\frac{1}{q^2/r_i} + \frac{1}{W_i n_i} = \frac{1}{P_E} \quad \text{or} \quad \frac{1}{P_0} = \frac{1}{q^2} + \frac{1}{(W r n)_i}, \quad P_E = P_0/r_i \qquad (3), (4), \text{ and } (5)$$

Here W_i = orbital energy of electrons [3], r_i = orbital radius of i orbital [4], $q = Z/n$ by [5, 6], n_i = number of electrons of the given orbital, Z and n = nucleus effective charge and effective main quantum number, and r = bond dimensional characteristics.

The value P_0 is called a spatial-energy parameter (SEP), and the value P_E = effective P–parameter (effective SEP). Effective SEP has a physical sense of some averaged energy of valence orbitals in the atom and is measured in energy units, for example, in electron volts (eV).

The values of P_0 parameter are tabulated constants for electrons of the given atom orbital.

For SEP dimensionality:

$$\left[P_0\right] = \left[q^2\right] = [E] \cdot [r] = [h] \cdot [\upsilon] = \frac{kgm^3}{s^2} = J\,m$$

where [E], [h], and [υ] = dimensionalities of energy, Planck's constant, and velocity.

The introduction of P-parameter should be considered as further development of quasi-classical concepts with quantum mechanical data on atom structure to obtain the criteria of phase formation energy conditions. For the systems of similarly charged (e.g. orbitals in the given atom) homogeneous systems the principle of algebraic addition of such parameters is preserved as follows:

$$\Sigma P_E = \Sigma \left(P_0 / r_i \right); \ \Sigma P_E = \frac{\Sigma P_0}{r} \qquad \text{(6) and (7)}$$

or:
$$\Sigma P_0 = P_0^{'} + P_0^{''} + P_0^{'''} + ...; \ r \Sigma P_E = \Sigma P_0 \qquad \text{(8) and (9)}$$

Here P-parameters are summed up by all atom valence orbitals.

To calculate the values of P_E-parameter at the given distance from the nucleus, either the atomic radius (R) or ionic radius (r_i) can be used instead of r depending on the bond type.

Let us briefly explain the reliability of such an approach. As the calculations demonstrated the values of P_E-parameters equal numerically (in the range of 2%), the total energy of valence electrons (U) by the atom statistic model. Using the known correlation between the electron density (b) and intra-atomic potential by the atom statistic model [7], we can obtain the direct dependence of P_E-parameter on the electron density at the distance r_i from the nucleus. The rationality of such technique was proved by the calculation of electron density using wave functions by Clementi [8] and comparing it with the value of electron density calculated through the value of P_E-parameter.

The modules of maximum values of the radial part of Ψ-function were correlated with the values of P_0-parameter and the linear dependence between these values was found. Using some properties of wave function as applicable to P-parameter, the wave equation of P-parameter with the formal analogy with the equation of Ψ-function was obtained [9].

8.2 WAVE PROPERTIES OF P-PARAMETERS AND PRINCIPLES OF THEIR ADDITION

Since P-parameter has wave properties (similar to Y′-function), the regularities of the interference of the corresponding waves should be mainly fulfilled at structural interactions.

The interference minimum, weakening of oscillations (in antiphase) occurs if the difference of wave move (Δ) equals the odd number of semi waves:

$$\Delta = (2n+1)\frac{\lambda}{2} = \lambda \left(n + \frac{1}{2} \right) \qquad (10)$$

where n = 0, 1, 2, 3, ...

As applicable to P-parameters this rule means that the interaction minimum occurs if P-parameters of interacting structures are also "in antiphase" either oppositely charged or heterogeneous atoms (for example, during the formation of valence-active radicals CH, CH_2, CH_3, NO_2, and so on) are interacting.

In this case P-parameters are summed by the principle of adding the reciprocals of P-parameters Equations (3 and 4).

The difference of wave move (Δ) for P-parameters can be evaluated via their relative value $\left(\gamma=\frac{P_2}{P_1}\right)$ of relative difference of P-parameters (coefficient a), which at the interaction minimum produce an odd number:

$$\gamma = \frac{P_2}{P_1} = \left(n+\frac{1}{2}\right) = \frac{3}{2};\frac{5}{2}...$$

$$\text{when } n = 0 \text{ (main state) } \quad \frac{P_2}{P_1} = \frac{1}{2} \quad (11)$$

It should be pointed out that for stationary levels of one-dimensional harmonic oscillator the energy of these levels $e = hn(n+\frac{1}{2})$, therefore, in quantum oscillator, in contrast to the classical one, the least possible energy value does not equal zero.

In this model the interaction minimum does not provide zero energy corresponding to the principle of adding reciprocals of P-parameters Equations (3 and 4).

The interference maximum and strengthening of oscillations (in phase) occurs if the difference of wave move equals the even number of semi waves:

$$\Delta = 2n\frac{\lambda}{2} = \lambda n \text{ or } \Delta = \lambda(n+1).$$

As applicable to P-parameters the maximum interaction intensification in the phase corresponds to the interactions of similarly charged systems or homogeneous systems by their properties and functions (e.g., between the fragments or blocks of complex inorganic structures, such as CH_2 and NNO_2 in octogene).

And then: $$\gamma = \frac{P_2}{P_1} = (n+1) \quad (12)$$

By the analogy, for "degenerated" systems (with similar values of functions) of two-dimensional harmonic oscillator the energy of stationary states:

$$e = hn(n+1)$$

By this model, the interaction maximum corresponds to the principle of algebraic addition of P-parameters – Equations (6 and 8). When n = 0 (main state) we have $P_2 = P_1$, or: the interaction maximum of structures occurs if their P-parameters are equal. This concept was used [2] as the main condition for isomorphic replacements and formation of stable systems.

8.3   EQUILIBRIUM EXCHANGE SPATIAL-ENERGY INTERACTIONS

During the formation of solid solutions and in other structural equilibrium exchange interactions the unified electron density should be established in the contact spots between atoms-components. This process is accompanied by the redistribution of electron density between valence areas of both particles and transition of a part of electrons from some external spheres into the neighboring ones.

It is obvious that with the proximity of electron densities in free atoms-components the transition processes between the boundaries atoms of particles will be minimal, thus contributing to the formation of a new structure. Thus, the task of evaluating the degree of such structural interactions in many cases comes down to comparative assessment of electron density of valence electrons in free atoms (on the averaged orbitals) participating in the process.

Therefore, the maximum total solubility evaluated *via* the structural interaction coefficient a is defined by the condition of minimal value of coefficient a, which represents the relative difference of effective energies of external orbitals of interacting subsystems as follows:

$$\alpha = \frac{P'_o / r_i' - P''_o / r_i''}{(P'_o / r_i' + P''_o / r_i'') / 2} 100\% \ \text{ or } \ \alpha = \frac{P'_s - P''_s}{P'_s + P''_s} 200\% \qquad \text{(13) and (14)}$$

where P_s = structural parameter is found by the following equation:

$$\frac{1}{P_s} = \frac{1}{N_1 \, P'_E} + \frac{1}{N_2 \, P''_E} + ... \qquad (15)$$

here N_1 and N_2 = number of homogeneous atoms in subsystems.

The nomogram of the dependence of structural interaction degree (ρ) upon the coefficient α, the same for the wide range of structures was prepared by the data obtained.

Isomorphism as a phenomenon is usually considered as applicable to crystalline structures. But the similar processes can also take place between molecular compounds where the bond energies can be accessed *via* the relative difference of electron densities of valence orbitals of interacting atoms. Therefore, the molecular electronegativity is rather easily calculated *via* the values of corresponding P-parameters.

In complex organic structures the main role in intermolecular and intramolecular interactions can be played by separate "blocks" or fragments considered as "active" areas of the structures. Therefore, it is necessary to identify these fragments and evaluate their SEPs. Based on wave properties of P-parameter, the total P-parameter of each element should be found following the principle of adding the reciprocals of initial P-parameters of all the atoms. The resulting P-parameter of the fragment block or all the structure is calculated following the rule of algebraic addition of P-parameters of their constituent fragments.

Apparently, spatial-energy exchange interactions (SEI) based on leveling the electron densities of valence orbitals of atoms-components have in nature the same universal value as purely electrostatic Coulomb interactions and complement each other. Isomorphism known from the time of E. Mitscherlich (1820) and D.I. Mendeleev (1856) is only a special demonstration of this general natural phenomenon.

The quantitative side of evaluating the isomorphic replacements both in complex and simple systems rationally fits into P-parameter methodology. More complicated is the problem of evaluating the degree of structural SEI for molecular structures, including organic ones. Such structures and their fragments are often not completely isomorphic to each other. Nevertheless, SEI is going on between them and its degree can be evaluated either semi-quantitatively numerically or qualitatively. By the degree of isomorphic similarity, all the systems can be divided into three types:

(I) Systems mainly isomorphic to each other—systems with approximately the same number of heterogeneous atoms and cumulatively similar geometric shapes of interacting orbitals.

(II) Systems with organic isomorphic similarity—systems, which,

> (1) either differs by the number of heterogeneous atoms but have cumulatively similar geometric shapes of interacting orbitals.
>
> (2) or have certain differences in the geometric shape of orbitals but have the same number of interacting heterogeneous atoms.

(III) Systems without isomorphic similarity—systems considerably different both by the number of heterogeneous atoms and geometric shape of their orbitals.

Taking into account the experimental data, all SEI types can be approximately classified as follows:

Systems I

1. $\alpha < (0–6)\%$; $\rho = 100\%$. Complete isomorphism, there is complete isomorphic replacement of atom-components.
2. $6\% < \alpha < (25–30)\%$; $\rho = 98 – (0–3)\%$.
   There is wide or unlimited isomorphism.
3. $\alpha > (25–30)\%$; no SEI.

Systems II

1. $\alpha < (0–6)\%$;
   (a) There is reconstruction of chemical bonds that can be accompanied by the formation of a new compound;
   (b) Cleavage of chemical bonds can be accompanied by a fragment separation from the initial structure but without adjoining and replacements.
2. $6\% < \alpha < (25–30)\%$; the limited internal reconstruction of chemical bonds is possible but without the formation of a new compound and replacements.
3. $\alpha > (20–30)\%$; no SEI.

Systems III

1. $\alpha < (0–6)\%$;
   (a) The limited change in the type of chemical bonds in the given fragment is possible, there is an internal regrouping of atoms without the cleavage from the main molecule part and replacements.
   (b) The change in some dimensional characteristics of the bond is possible.
2. $6\% < \alpha < (25–30)\%$;
   A very limited internal regrouping of atoms is possible;
3. $\alpha > (25–30)\%$; no SEI.

When considering the above systems, it should be pointed out that they can be found in all cellular and tissue structures in some form but are not isolated and are found in spatial time combinations.

The values of α and ρ calculated in such a way refer to a definite interaction type whose monogram can be specified by fixed points of reference systems. If we take into account the universality of spatial-energy interactions in nature, such evaluation can have the significant meaning for the analysis of structural shifts in complex biophysical and chemical processes of biological systems.

Fermentative systems contribute a lot to the correlation of structural interaction degree. In this model, the ferment role comes to the fact that active parts of its structure (fragments, atoms, ions) have such a value of P_E-parameter, which equals the P_E-parameter of the reaction final product. That is the ferment is structurally "tuned" *via* SEI to obtain the reaction final product, but it will not enter it due to the imperfect isomorphism of its structure (in accordance with III).

The important characteristics of atom-structural interactions (mutual solubility of components, chemical bond energy, energy of free radicals, and so on) for many systems were evaluated following this technique [10, 11].

8.4 CALCULATION OF INITIAL DATA AND BOND ENERGIES

Based on the Equations (3-5) with the initial data calculated by quantum mechanical methods [3-6], we calculate the values of P_0-parameters for the majority of elements being tabulated and constant values for each atom valence orbital. Mainly covalent radii by the main type of the chemical bond of interaction considered were used as a dimensional characteristic for calculating P_E-parameter (Table 1). The value of Bohr radius and the value of atomic ("metal") radius were also used for hydrogen atom.

In some cases, the bond repetition factor for carbon and oxygen atoms was taken into consideration [10]. For a number of elements, the values of P_E-parameters were calculated using the ionic radii whose values are indicated in column 7. All the values of atomic, covalent, and ionic radii were mainly taken by Belov-Bokiy and crystalline ionic radii by Batsanov [12].

The results of calculating structural P_S-parameters of free radicals by the Equation (15) are given in Table 2. The calculations are done for the radicals contained in protein and amino acid molecules (CH, CH_2, CH_3, NH_2, and so on), as well as for some free radicals formed in the process of radiolysis and dissociation of water molecules.

The technique previously tested [10] on 68 binary and more complex compounds was applied to calculate the energy of coupled bond of molecules by the following equations:

$$\frac{1}{E} = \frac{1}{P_S} = \frac{1}{\left(P_E \dfrac{n}{K}\right)_1} + \frac{1}{\left(P_E \dfrac{n}{K}\right)_2} \; ; \; P_E \frac{n}{K} = P \qquad \text{(16) and (17)}$$

where n = bond average repetition factor and K = hybridization coefficient, which usually equals the number of registered atom valence electrons.

TABLE 1 P-parameters of atoms calculated via the bond energy of electrons.

Atom	Valence electrons	W (eV)	r_i (Å)	q^2_0 (eVÅ)	P_0 (eVÅ)	R (Å)	P_0/R (eV)
H	$1S^1$	13.595	0.5295	14.394	4.7985	0.5292	9.0644
						0.28	17.137
						$R'_i = 1.36$	3.525
C	$2P^1$	11.792	0.596	35.395	5.8680	0.77	7.6208
						0.67	8.7582
	$2P^2$	11.792	0.596	35.395	10.061	0.77	13.066
						0.67	15.016
	$2S^2$				14.524	0.77	18.862
	$2S^2+2P^2$				24.585	0.77	31.929
					24.585	0.67	36.694
N	$2P^1$	15.445	0.4875	52.912	6.5916	0.70	9.4166
	$2P^2$				11.723	0.70	16.747
	$2P^3$				15.830	0.70	22.614
						0.55	28.782
	$2S^2$	25.724	0.521	53.283	17.833	0.70	25.476
	$2S^2+2P^3$				33.663	0.70	48.09

TABLE 1 *(Continued)*

Atom	Valence electrons	W (eV)	r_i (Å)	q^2_0 (eVÅ)	P_0 (eVÅ)	R (Å)	P_0/R (eV)
O	$2P^1$	17.195	0.4135	71.383	6.4663	0.66	9.7979
	$2P^1$					$R_1 = 1.36$	4.755
	$2P^1$					$R_1 = 1.40$	4.6188
	$2P^2$	17.195		71.383	11.858	0.66	17.967
						0.59	20.048
						$R_1 = 1.36$	8.7191
						$R_1 = 1.40$	8.470
	$2P^4$	17.195	0.4135	71.383	20.338	0.66	30.815
						0.59	34.471
	$2S^2$	33.859	0.450	72.620	21.466	0.66	32.524
	$2S^2+2P^4$				41.804	0.66	63.339
						0.59	70.854
Ca	$4S^1$	5.3212	1.690	17.406	5.929	1.97	3.0096
	$4S^2$				8.8456	1.97	4.4902
	$4S^2$					$R^{2+} = 1.00$	8.8456
	$4S^2$					$R^{2+} = 1.26$	7.0203
P	$3P^1$	10.659	0.9175	38.199	7.7864	1.10	7.0785
	$3P^1$					$R^{3+} = 1.86$	$P_3 = 4.1862$
	$3P^3$	10.659	0.9175	38.199	16.594	1.10	15.085
	$3P^3$					$R^{3+} = 1.86$	8.9215
	$3S^2+3P^3$				35.644	1.10	32.403

TABLE 1 (*Continued*)

Atom	Valence electrons	W (eV)	r_i (Å)	q^2_0 (eVÅ)	P_0 (eVÅ)	R (Å)	P_0/R (eV)
Mg	$3S^1$	6.8859	1.279	17.501	5.8568	1.60	3.6618
	$3S^2$				8.7787	1.60	5.4867
						$R^{2+} = 1.02$	8.6066
Mn	$4S^1$	6.7451	1.278	25.118	6.4180	1.30	4.9369
	$4S^1+3d^1$				12.924	1.30	9.9414
	$4S^2+3d^2$				22.774	1.30	17.518
Na	$3S^1$	4.9552	1.713	10.058	4.6034	1.89	2.4357
						$R^{1+}_i = 1.18$	3.901
						$R^{1+}_i = 0.98$	4.6973
K	$4S^1$	4.0130	2.612	10.993	4.8490	2.36	2.0547
						$R^{1+}_i = 1.45$	3.344

Here, the P-parameter of energy characteristic of the given component structural interaction in the process of binary bond formation.

"Non-valence, non-chemical weak forces act … inside biological molecules and between them apart from strong interactions [13]." At the same time, the orientation, induction, and dispersion interactions are used to be called Van der Waals. For three main biological atoms (nitrogen, phosphorus, and oxygen), Van der Waals radii numerically equal approximately the corresponding ionic radii (Table 3).

It is known that one of the reasons of relative instability of phosphorus anhydrite bonds in ATP is the strong repulsion of negatively charged oxygen atoms. Therefore, it is advisable to use the values of P-parameters calculated *via* Van der Waals radii as the energy characteristic of weak structural interactions of biomolecules (Table 3).

TABLE 2 Structural P_S-parameters calculated *via* the bond energy of electrons.

Radicals, molecule fragments	$P_i^{'}(eV)$	$P_i^{''}(eV)$	Ps(eV)	Orbitals
OH	9.7979	9.0644	4.7080	O $(2P^1)$
	17.967	17.138	8.7712	O $(2P^2)$
H_2O	2·9.0644	17.967	9.0227	O $(2P^2)$
CH_2	17.160	2·9.0644	8.8156	C $(2S^{1}2P^{3}_{r})$
	31.929	2·17.138	16.528	C $(2S^{2}2P^{2})$
CH_3	15.016	3·9.0644	9.6740	C $(2P^2)$
	40.975	3·9.0644	16.345	C $(2S^{2}2P^{2})$
CH	31.929	12.792	9.1330	C $(2S^{2}2P^{2})$
NH	16.747	17.138	8.4687	N$(2P^2)$
	19.538	17.132	9.1281	N$(2P^2)$
NH_2	19.538	2·9.0644	9.4036	N$(2P^2)$
	28.782	2·17.132	18.450	N$(2P^3)$
CO–OH	8.4405	8.7710	4.3013	C$(2P^2)$
C=O	15.016	20.048	8.4405	C$(2P^2)$
C=O	31.929	34.471	16.576	O$(2P^4)$
CO=O	36.694	34.471	17.775	O$(2P^4)$
C–CH$_3$	17.435	19.694	9.2479	–
C–NH$_2$	17.435	18.450	8.8844	–

TABLE 2   *(Continued)*

Radicals, molecule fragments	$P_i^{'}(eV)$	$P_i^{*}(eV)$	Ps(eV)	Orbitals
CO–OH	12.315	8.7712	5.1226	$C(2S^22P^2)$
(HP)O$_3$	23.122	23.716	11.708	$O(2P^2)$ $P(3S^23P^3)$
(H$_3$P)O$_4$	17.185	17.244	8.6072	$O(2P^1)$ $P(3P^1)$
(H$_3$P)O$_4$	31.847	31.612	15.865	$O(2P^2)$ $P(3S^23P^3)$
H$_2$O	2·4.3623	8.7191	4.3609	$O(2P^2)$ $r = 1.36$ Å
H$_2$O	2·4.3623	4.2350	2.8511	$O(2P^2)$ $r = 1.40$
C-H$_2$O	2.959	2.8511	1.4520	–
(C-H$_2$O)$_3$ Lactic acid	–	–	1.4520·3= 4.3563	–
(C-H$_2$O)$_6$ Glucose	–	–	1.4520·6= 8.7121	–

Bond energies for P and O atoms were calculated taking into account Van der Waals distances for atomic orbitals: $3P^1$ (phosphorus)-$2P^1$ (oxygen) and for $3P^3$ (phosphorus)-$2P^2$ (oxygen). The values of E obtained slightly exceeded the experimental Reference ones (Table 4). But for the actual energy physiological processes, for example during photosynthesis, the efficiency is below the theoretical one, being about 83%, in some cases [14, 15].

Perhaps the electrostatic component of resulting interactions at anion-anionic distances is considered in such a way that the calculated value of 0.83 E practically corresponds to the experimental values of bond energy during the phospholyration and free energy of ATP in chloroplasts.

Table 4 contains the calculations of bond energy following the same technique but for stronger interactions at covalent distances of atoms for the free molecule P=O (sesquialteral bond) and for the molecule P = O (double bond).

The sesquialteral bond was evaluated by introducing the coefficient $n = 1.5$ with the average value of oxygen P_E-parameter for single and double bonds.

The average breaking energy of the corresponding chemical bonds in ATP molecule obtained in the frameworks of semi-empirical method PM3 with the help of software GAMESS [16] are given in column 11 of Table 4 for comparison. The calculation technique is detailed by Khokhriakov et al. [17].

The calculated values of bond energies in the system K-C-N being close to the values of high-energy bond P~O in ATP demonstrate that such structure can prevent the ATP synthesis.

When evaluating the possibility of hydrogen bond formation, we take into account such value of n/K in which K = 1 and the value $n = 3.525/17.037$ characterizes the change in the bond repetition factor when transiting from the covalent bond to the ionic one.

8.5   FORMATION OF STABLE BIOSTRUCTURES

At equilibrium exchange spatial-energy interactions similar to isomorphism, the electrically neutral components do not repulse but approach each other and form a new composition whose α is present in the Equations (13) and (14).

This is the first stage of stable system formation by the given interaction type, which is carried out under the condition of approximate equality of component P-parameters: $P_1 \approx P_2$.

Hydrogen atom element No 1 with the orbital $1S^1$ determines the main criteria of possible structural interactions. Four main values of its P-parameters can be taken from Tables 1 and 3.

(1) For strong interactions: $P_E' = 9.0644$ eV with the orbital radius 0.5292 Å and $P_E'' = 17.137$ eV with the covalent radius 0.28 Å.

(2) For weaker interactions: $P_E' = 4.3623$ eV and $P_E = 3.6352$ eV with Van der Waals radii 1.10 Å and 1.32 Å. The values of P-parameters $P' : P' : P''$ relates as 1:2:4. Such values of interaction P-parameters define the normative functional states of biosystem and the intermediary can produce pathologic formations by their values.

The series with approximately similar values of P-parameters of atoms or radicals can be extracted from the large pool of possible combinations of structural interactions (Table 5). The deviations from the initial and primary values of P-parameters of hydrogen atom are in the range ±7%.

The values of P-parameters of atoms and radicals given in the Table define their approximate equality in the directions of interatomic bonds in polypeptide, polymeric, and other multi-atom biological systems.

In ATP molecule, these are phosphorus, oxygen and carbon atoms, and polypeptide chains –CO, NH, and CH radicals. In Table 5, the additional calculation of their bond energy taking into account the sesquialteral bond repetition factor in radicals C~O и N~H can be also seen.

On the example of phosphorus acids, it can be demonstrated that this approach is not in contradiction with the method of valence bonds, which explains the formation peculiarities of ordinary chemical compounds. It is demonstrated in Table 6 that this electrostatic equilibrium between the oppositely charged components of these acids can correspond to the structural interaction for H_3PO_4 $3P^1$ orbitals of phosphorus and $2P^1$ of oxygen, and for HPO_3 $3S^23P^3$ orbitals of phosphorus and $2P^2$ of oxygen. Here, it is stated that P-parameters for phosphorus and hydrogen subsystems are added algebraically. It is also known that the ionized phosphate groups are transferred in the process of ATP formation that is apparently defined for phosphorus atoms by the transition from valence active $3P^1$ orbital to $3S^23P^3$ ones that is 4 additional electrons will become valence active. According to the experimental data, the synthesis of one ATP molecule is connected with the transition of four protons and when the fourth proton is being transited the energy accumulated by the ferment reaches its threshold [18, 19]. It can be assumed that such proton transitions in ferments initiate similar changes in valence active states in the system P-O. In the process of oxidating pholpholyration, the transporting ATP-syntheses use the energy of gradient potential due to $2H^+$-protons. In the given model, such a process corresponds to the initiation of valence active transitions of phosphorus atoms from $3P^1$ to $3P^3$ state.

In accordance with the Equation (17), we can assume that in stable molecular structures the condition of the equality of corresponding effective interaction energies of the components by the couple bond line is fulfilled by the following equations:

$$\left(P_E \frac{n}{K} \right)_1 \approx \left(P_E \frac{n}{K} \right)_2 \rightarrow P_1 \approx P_2 \qquad (18)$$

And for heterogeneous atoms (when $n_1 = n_2$):

$$\left(P_E \Big/ K \right)_1 \approx \left(P_E \Big/ K \right)_2 \qquad (18(a))$$

In phosphate groups of ATP molecule, the bond main line comprises phosphorus and oxygen molecules. The effective energies of these atoms by the bond line calculated by the Equation (18) are given in Tables 4 and 5 from which it is seen that the best equality of P_1 and P_2 parameters is fulfilled for the interactions $P(3P^3) - 8.7337$ eV and $O(2P^2) - 8.470$ eV that is defined by the transition from the covalent bond to Van der Waals ones in these structures.

The resulting bond energy of the system P–O for such valence orbital and the weakest interactions (maximum values of coefficient K) is 0.781 eV (Table 4). Similar calculations for the interactions $P(2P^1) - 4.0981$ eV and $O(2P^1) - 4.6188$ eV produce the resulting bond energy 0.397 eV.

The difference in these values of bond energies is defined by different functional states of phosphorous acids HPO_3 and H_3PO_4 in glycolysis processes and equals 0.384 eV that is close to the pholpholyration value (0.340.35 eV) obtained experimentally.

Such ATP synthesis is carried out in anaerobic conditions and is based on the transfer of phosphate residues onto ATP *via* the metabolite. For example, ATP formation from creatine phosphate is accompanied by the transition of its NH group at ADP to NH_2 group of creatine at ATP.

TABLE 3 Ionic and Van der Waals radii (Å).

Atom	Ionic radii			Van der Waals radii		
	Orbital	R_I	P_E/κ (eV)	R_B	Orbital	P_E/κ (eV)
H	$1S^1$	$R^- = 1.36$	3.525	1.10	$1S^1$	4.3623
		$r = 0.5292$	9.0644	1.32		3.6352
N	$2P^3$	$R^{3-} = 1.48$	$10.696/3 = 3.5653$	1.50	$2P^1$	4.3944/1
				1.50	$2P^3$	$10.553/3 = 3.5178$
				1.50	$2S^2 2P^3$	$22.442/5 = 4.4884$
P	$3P^3$	$R^{3-} = 1.86$	$8.9215/3 = 2.9738$	1.9	$3P^1$	4.0981/1
				1.9	$3P^3$	$8.7337/3 = 2.9112$
		$R^{2-} = 1.40$	$8.470/2 = 4.2350$	1.9	$3S^2 3P^3$	$18.760/5 = 3.752$
				1.40	$2P^1$	4.6188/1
				1.50	$2P^1$	4.3109/1
O	$2P^2$	$R^{2-} = 1.36$	$8.7191/2 = 4.3596$	1.40	$2P^2$	$8.470/2 = 4.2350$
				1.50	$2P^2$	$7.9053/2 = 3.9527$
				1.7	$2P^1$	3.4518/1
C	$2S^2 2P^2$	$d^*/2 = 3.2/2 = 1.6$	$15.365/4 = 3.841$	1.7	$2P^2$	$5.9182/2 = 2.9591$
				1.7	$2S^2 2P^2$	$14.462/4 = 3.6154$

d = contact distance between C–C atoms in polypeptide chains [13].

TABLE 4 Bond energy (eV).

Atoms, structures, and orbitals	Bond	Component 1		Component 2		Component 3		Calculation	E [1315]	E [16, 17]	Remarks
		P_E (eV)	n/K	P_E (eV)	n/K	P_E (eV)	n/K	E			
1	2	3	4	5	6	7	8	9	10	11	12
P–O 3S²3P¹–2S²2P⁴	cov.	32.403	1.5/5	70.854	1.5/6	6.14		6.277	6.1385		PO free molecule
				63.339	1.5/6			6.024 <6.15>	6.14		
H₂O	cov.	2×9.0624	1/1	17.967	1/6			2.570	2.476		Decay
1S¹–2P²	cov.	2×9.0624	1/1	20.048	2/2			9.520		10.04	of one molecule
H₃PO₄	cov.	3×9.0624	1/1	32.405	1/5	4×17.967	1/2	4.8779	4.708		
C–O (2P¹–1S¹)	cov.	7.6208	1.125/2	9.7979	1/1			4.2867			
C–N 2P¹–2P¹	cov.	7.6208	1/4	9.4166	1/5			0.9471			
(C–H₂O)–(C–H₂O)	VdW	1.4520	1/1	1.4520	1/1			0.726			
C–O	cov.	31.929	1.125/4	20.048	1/2			4.7367			
2S²2P²–2P²		31.929	1/4	20.042	1/2			4.4437			
N–H	cov.	9.4166	1.1667/1	9.0644	1/1			4.9654			
2P¹–1S¹		9.4166	1/1	9.0644	1/1			4.6186			
C–H 2P¹–1S¹		13.066	1/2	9.0644	1/1			3.797	3.772		
C–H 2P¹–1S¹	cov.	13.066	1/2	17.137	1/1			4.7295			
N–H₂ 2P¹–1S¹	cov.	22.614	1/3	2×9.0644	1/1			5.3238			
–H···O		3.525	3.525/ 17.037	4.6188	1/6			0.3730	0.3742		Hydrogen bond

TABLE 4 *(Continued)*

								Free molecule
P=O 3P¹-2P²	cov.	15.085	2/3	20.042	2/2	6.6970	6.504	6.1385
P-O 3P¹-2P²	VdW	8.7337	1/5	8.470	1/6	0.781	0.670	
P-O 3P¹-2P¹	cov.	7.0785	1/1	9.7979	1/1	4.1096	4.2059	4.2931
P-O 3P¹-2P¹	VdW	4.0981	1/5	4.6188	1/6	0.3970	0.34-0.35	

ΔG ATP

Phospho lyration

TABLE 5 Bio-structural spatial-energy parameters (eV).

Series No	H	C	N	O	P	CH	CO	NH	Glucose	Lactic acid	OH	Remarks
I	9.0644 (1S¹)	8.7582 (2P¹) 9.780 (2P¹)	9.4166 (2P¹)	9.7979 (2P¹)	8.7337 (3P³)	9.1330 (2S²2P²–1S¹)	8.4405 (2P²–2P²)	8.4687 (2P²–1S¹) 9.1281 (2P²–1S¹)	8.7121 2P²– –(1S¹–2P²)		8.7710	Strong interaction
II	17.132 (1S¹)	17.435 (2S¹2P¹)	16.747 (2P²)	17.967 (2P²)	18.760 (3S³3P³)	C and H blocks	16.576 (2S²2P²–2P⁴)	N and H blocks				Strong interaction
III	(4.3623) (1S¹)	3.8696 (2P²)	4.3944 (2P¹)	4.3109 (2P¹) 4.6188 (2P¹)	4.0981 (3P¹)	4.7295	4.4437 4.7367	4.6186 4.9654		4.3563 2P²– (1S¹–2P²)	4.7084	Weak interaction
IV	3.6352 (1S¹)	3.4518 (2P¹) 3.6154 (2S²2P²)	3.5178 (2P³)	4.2350 (2P²) 3.6318 (2P⁴)	4.0981 (3P¹) 3.752 (3S²3P³)	4.7295	4.4437 4.7367	4.6186 4.9654				Effective bond energy

TABLE 6: Structural interactions in phosphorus acids.

Molecule	Component 1			Component 2			$\alpha = (\Delta P / \langle P \rangle) * 100\%$
	Atom	Orbitals	$P = P_1 + P_2$ (eV)	Atom	Orbit-als	P(eV)	
$(H_3P)O_4$	H_3P	$1S^1 - 3P^1$	$4.3623*3 + 4.0981 = 17.185$	O_4	$2P^1$	$4.3109*4 = 17.244$	0.34
		$1S^1 - (3S^2 3P^2)$	$4.3623*3 + 18.760 = 31.847$	at $r = 1.50\text{Å}$	$2P^2$	$7.9053*4 = 31.612$	0.74
$(HP)O_3$	HP	$1S^1 - (3S^2 3P^2)$	$4.3623 + 18.760 = 23.122$	O_3	$2P^2$	$7.9053*3 = 23.716$	2.54
				at $r = 1.50\text{Å}$			

From Table 4, it is seen that the change in the bond energy of these two main radicals of metabolite is 5.3238 – 4.9654 = 0.3584 eV, taking the sesquialteral bond N--H into account (as in polypeptides) and 5.3238 – 4.6186 = 0.7052 eV for the single bond N-H. This is one of the intermediary results of the high-energy bond transformation process in ATP through the metabolite. From Tables 4 and 6, we can conclude that the phosphorous acid H_3PO_4 can have two stationary valence active states during the interactions in the system P-O for the orbital with the values of P-parameters of weak and strong interactions, respectively. This defines the possibility for the glycolysis process to flow in two stages. At the first stage, the glucose and H_3PO_4 molecules approach each other due to similar values of their P-parameters of strong interactions (Table 2). At the second stage, H_3PO_4 P-parameter in weak interactions 4.8779 eV (Table 4) in the presence of ferments provokes the bond $(H_2O-C)-(C-H_2O)$ breakage in the glucose molecule with the formation of two molecules of lactic acid whose P-parameters are equal by 4.3563 eV. The energy of this bond breakage process equalled to 0.726 eV (Table 4) is realized as the energy of high-energy bond it ATP.

According to the reference data, about 40% of the glycolysis total energy, that is about 0.83 eV, remains in ATP.

By the hydrolysis reaction in ATP in the presence of ferments

$(HPO_3 + H_2O \rightarrow H_3PO_4 + E)$ for structural P_S-parameters (Table 2) E = 11.708 + 4.3609 15.865 = 0.276 eV

It is known that the change in the free energy (ΔG) of hydrolysis of phosphorous anhydrite bond of ATP at pH = 7 under standard conditions is 0.3110.363 eV. But in the cell, the ΔG value can be much higher as the ATP and ADP concentration in it is lower than under standard conditions. Besides, the ΔG value is influenced by the concentration of magnesium ions, which is the acting co-ferment in the complex with ATP. Actually, Mg^{2-} ion has the P_E-parameter equalled to 8.6066 eV (Table 1), which is very similar to the corresponding values of P-parameters of phosphorous and oxygen atoms.

The quantitative evaluation of this factor requires additional calculations.

8.6 CONCLUSION

1. The bond energies of some biostructures have been calculated following P-parameter and quantum mechanical techniques.
2. The high-energy bonds in ATP are formed in the system P-O under functional transitions of their valence active states.
3. The data obtained agree with the experimental ones.

KEYWORDS

- **Heterogeneous systems**
- **Isomorphism**
- **Lagrangian equation**
- **Potential energy**
- **Spatial-energy P-parameter**

REFERENCES

1. Eyring, H., Walter, J., and Kimball, G. E. *Quantum chemistry I. L., M.*, p. 528 (1948).
2. Korablev, G. A. *Spatial-Energy Principles of Complex Structures Formation*. Brill Academic Publishers and VSP, Netherlands, p. 426. (Monograph) (2005).
3. Fischer, C. F. Average-Energy of Configuration Hartree-Fock Results for the Atoms Helium to Radon. *Atomic Data*, 4(4), 301399 (1972).
4. Waber, J. T. and Cromer, D. T. Orbital Radii of Atoms and Ions. *J. Chem. Phys.*, 42(12), 41164123 (1965).
5. Clementi, E. and Raimondi, D. L. Atomic Screening constants from S. C. F. Functions, 1. *J. Chem. Phys.*, 38(11), 26862689 (1963).
6. Clementi, E. and Raimondi D. L. Atomic Screening constants from S. C. F. Functions, 1. *J. Chem. Phys.*, 47(14), 13001307 (1967).
7. Gombash, P. Atom statistical model and its application. M.I.L., p. 398 (1951).
8. Clementi, E. Tables of atomic functions. *J. B. M. S. Re. Develop. Suppl.*, 9(2), 76 (1965).
9. Korablev, G. A. and Zaikov, G. E. Spatial-Energy Parameter as a Materialised Analog of Wafe Function. Progress on Chemistry and Biochemistry, Nova Science Publishers, Inc. New York, 355376 (2009).
10. Korablev, G. A. And Zaikov, G. E. Energy of chemical bond and spatial-energy principles of hybridization of atom orbitals. *J. of Applied Polymer Science*, 101(3), 21012107 (Aug 5, 2006).
11. Korablev, G. A. and Zaikov, G. E. Formation of carbon nanostructures and spatial-energy criterion of stabilization. *Mechanics of composite materials and structures, RAS*, 15(1), 106118 (2009).
12. Batsanov, S. S. *Structural chemistry. Facts and dependencies*. M. MSU, (2009).
13. Volkenshtein. *Biophysics. M.: Nauka, 598 (1988).*
14. Photosynthesis (Ed.) by Govindzhi. M.: Mir, 1, 2, 728, 460 (1987).
15. Clayton, R. Photosynthesis. *Physical mechanisms and chemical models. M.: Mir,* 350 (1984).
16. Schmidt, M. W., Baldridge, K. K., and Boatz, J. A. et al. General atomic and molecular electronic structure system. *J. Comput. Chem.*, 14, 13471363 (1993).
17. Khokhriakov, N. V. and Kodolov, V. I. Influence of active nanoparticles on the structure of polar liquids. *Chemical physics and mesoscopy*, 11(3), 388402 (2009).
18. Feniouk, B. A. Study of the conjugation mechanism of ATP synthesis and ATP proton transport. Referun - *Biology and natural science*, 108 (1998).
19. Feniouk, B. A., Junge, W., and Mulkidjanian, A. Tracking of proton flow across the active ATP-synthase of Rhodobacter capsulatus in response to a series of light flashes. *EBEC Reports*, 10, 112, (1998).

9 Temperature Dependence of Self-adhesion Bonding Strength for Miscible Polymers

Kh. Sh. Yakh'yaeva, G. V. Kozlov,
G. M. Magomedov, and G. E. Zaikov

CONTENTS

9.1   INTRODUCTION

The quantitative structural model of self-adhesion bonding strength for miscible amorphous polymers was described showing a good correspondence to the experiment. This model allows defining factors affecting the self-bonding strength. At present, it has been generally acknowledged that in self-adhesion bonding, the mechanical properties depend on two factors, namely: The macromolecular entanglements formation in the boundary layer and the macromolecular coils interdiffusion kinetics [1]. However, as a rule, an experimental results treatment of self bonded layers may give rise to only qualitative analysis. Instead, the present communication purpose is the development of quantitative structural model for self-adhesion bonding strength, using the fractal analysis notions.

9.2   EXPERIMENTAL

Amorphous polystyrene (PS, $M_w = 23 \times 10^4$, $M_w/M_n = 2.84$) and poly (2,6-dimethyl 1,4-phenylene oxide) (PPO, $M_w = 44 \times 10^3$, $M_w/M_n = 1.91$), obtained from Dow Chemical and General Electric (USA), respectively, were used. The films of polymers with thickness about 100 microns were prepared by an extrusion method. The PS-PPO interfaces were healed during 10 min within the temperature range of 343–396K at pressure of 0.8 MPa. The lap shear strength of bonded polymers was conducted at temperature 293K by means of an Instron tensile tester, model 1130 at a crosshead speed of 3 ′ 10⁻² m/s.

9.3   DISCUSSION AND RESULTS

In Figure 1, the experimental dependence of shear strength, τ_b, of PS-PPO bonding on the bonding temperature is shown. The authors [1] explained the indicated dependence by faster macromolecular coils interdiffusion at increasing temperature. In a previous work, the shear strength, τ_b, was found to obey the following equation:

$$\ln \tau_b = N_i - 16.6D_f + 20.5 \qquad (1)$$

where N_i is the number of coils intersections along the boundary layer, physically describing the macromolecular entanglements density, D_f is the macromolecular coil dimension.

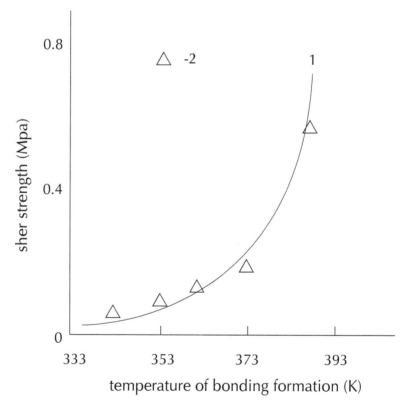

FIGURE 1   The dependences of shear strength τb of self-adhesion bonding PS-PPO as function of the bonding formation temperature T. Line: the calculation according to the Equation (1). Symbols: experimental data [1].

The Equation (1) takes into consideration both the factors influencing the shear strength, τ_b, namely: the number of coils intersections, N_i that takes into account the macromolecular entanglements formation, and the coil dimension ($-16.6\,D_f$) that accounts for the interdiffusion weakening at increasing D_f.

N_i can be determined according to the following fractal relationship [2]:

$$N_i \sim R_g^{D_{f1}+D_{f2}-d} \tag{2}$$

where R_g is macromolecular coil gyration radius, D_{f_1} and D_{f_2} are fractal dimension of coils structure, forming self-adhesion bonding, d is the dimension of Euclidean space, in which a fractal is considered (it is obvious that in our case $d = 3$).

For the estimation of the parameter D_f, the following approximated technique will be used. It is known that the correlation between D_f and structure dimension, d_f of linear polymers in the condensed state obeys the following equation:

$$D_f = \frac{d_f}{1.5} \tag{3}$$

With the d_f estimation conducted according to the following formula [3]:

$$d_f = 3 - 6\left(\frac{\varphi_{cl}}{SC_\infty}\right)^{1/2} \tag{4}$$

where φ_{cl} is relative fraction of local order domains (clusters), S is the macromolecule cross section area, and C_∞ is a characteristic ratio.

The estimation of φ_{cl} was performed according to the following percolation relationship [3]:

$$\varphi_{cl} = 0.03\left(T_g - T\right)^{0.55} \tag{5}$$

where T_g and T are glass transition and the bonding temperatures, respectively.

It was already found that for PS $C_\infty = 9.8$ and $S = 54.8$ Å^2 and for PPO $C_\infty = 3.8$ and $S = 27.9$ Å^2 [3]. The macromolecular coil gyration radius was calculated further as follows [3–6]:

$$R_g = l_0\left(\frac{C_\infty M_w}{6m_0}\right)^{1/2} \tag{6}$$

where l_0 is length of the main chain skeletal bond, which is equal to 0.154 nm for PS and 0.541 nm for PPO, m_0 is the molar mass per backbone bond ($m_0 = 52$ for PS and $m_0 = 25$ for PPO [4]).

Let us note an important methodological aspect. The R_g was calculated according to the Equation (6) where $C_\ast$ was considered a variable that can be calculated according to the following relationship [3]:

$$\tilde{N}_\infty = \frac{2d_f}{d(d-1)(d-d_f)} + \frac{4}{3}$$

(7)

In Figure 1, the comparison of experimental data and the theoretical curve obtained on the basis of the indicated method (the solid curve) is reported. Based on Equation (1), the $\tau_b(T)$ calculation gives rise to the following considerations. First, the mean R_g value for PS and PPO was used as macromolecular coil gyration radius. Second, at $T = 386$ K, where $T > T_g$ for PS, the value $d_f = 2.95$, that is the greatest one for real solids [3], was used to calculate $C_\ast$ according to the equation (7). As it follows from the data of Figure 1, the good correspondence of theory and experiment was obtained that confirms the technique reliability.

Hence, from the analysis of the results, the shear strength, τ_b, is influenced by a number of factors, namely: the glass transition T_g of polymers and the bonding temperature T, the molecular characteristics, $C_\ast$, S, l_0, m_0, and the molecular weight of polymers M_w.

KEYWORDS

- **Autohesion**
- **Fractal analysis**
- **Intersection**
- **Macromolecular coil**
- **Miscible polymers**

REFERENCES

1. Boiko, Yu. M. and Prud'homme, R. E. *J. Polymer Sci.: Part B: Polymer Phys.*, **36**, 567–572 (1998).
2. Vilgis, T. A. *Physica A*, **153**, 341–354 (1988).
3. Kozlov, G. V. and Zaikov, G. E. "*Structure of the Polymer Amorphous State*". Boston, Brill Academic Publishers, Leiden, p. 465 (2004).
4. Grassia, L. and D'Amore, A., *Journal of Rheology*, **53**(2), 339–356 (2009).
5. Schnell, R., Stamm, M., and Creton, C. *Macromolecules*, **31**, 2284–2292 (1998).
6. Basile, A., Greco, F., Mader, A., and Carrà, S., *Plastics, Rubber and Composites*, **32**(8–9), 340–344 (2003).

10   A Note on Biological Damage and Protection of Materials

Elena L. Pekhtasheva, Anatoly N. Neverov,
Stefan Kubica, and Gennady E. Zaikov

CONTENTS

10.1   INTRODUCTION

A healthy growing tree stem is not usually damaged by insects. Osmotic pressure in tissues maintained at a definite level in the course of water supply through the roots into the tree stem prevents their appearance. As soon as normal water exchange is disturbed and osmotic pressure in the whole stem or its part changes, the tree will be exposed to insect attack [1].

Insects attacking the tree stem have very well developed mouthparts with two couples of jaws, which help insects gnawing easily through the rind and dispersing the tree tissues. After cutting, the tree loses its protective properties. Due to chemical processes proceeding in it, the fallen tree has a strong characteristic smell, which attracts insects. They find fallen trees very quickly and colonize them. Exposed to insects, the wood loses its eligibilities very soon, durability decreases, endurance and bearing strength distinctly abate, wetness, bulk, and thermal conductivity change. Fungus spores easily penetrate into this wood and it begins to rot. Within some years, wood becomes rotten, decays to pieces, and becomes easy powdered in hands. If such wood is being used as a post, floor beam, it will not endure the necessary load [2].

Wood pests can get from forest to multistory house with the tree. Usually, if an insect is already delivered to the house from the forest, it cannot immediately accommodate itself to the life in house and dies. However, there are pest species in nature, which over a long period have being accommodating to the life in old ram pike, where they passed into the human being house from. The wood fretter refers to such pests [3, 4].

10.2 CURIOUS FACTS

The majority of their life mature wood fretters spend in burrows in the ram pike, where usually their larvae are developing. Many species produce sounds by tapping burrow´s walls with the head. They do it so rhythmically that it has an impression of clock ticking. Superstitious people call them clock of death and believe that this is a bad presage. In fact, these sounds help beetle females and males to find each other in the wood strata. When the clock ticking sounds from a wall, table or cupboard, this is a warning for the house owner that his house or furniture is being damaged by hazardous enemies the wood fretters.

Common furniture beetle refers to a small fretter family, which includes about 200 species. The majority of them are met in the warm climate [5, 6].

In the course of historical development, the connection between expansion of common furniture beetle and human business activities is strengthening ever more. Common furniture beetle is generally found only in buildings. This is because common furniture beetle is transported with infested wood and articles from it, and in buildings, it usually finds suitable conditions for development.

The wood fretter larvae develop in the inner parts of woodworks frequently making them rot, whereas from the outside only relatively small circular exit holes, where beetles leave woodworks through are visible. Their larvae develop up to 3–4 years, so even an expert cannot detect infected components at once, but only after the first beetle flight. Most often wood fretters appear in premises together with the old woodwork, the second way is the beetle flight into open windows.

Common furniture beetle scathe cannot always be exactly evaluated, because wood is usually dispersed not only by the beetle, but also by wood fungi. So far wood fretter often destroys not only usual woodworks and buildings, but also unique ones of historical and museum value, it becomes impossible to evaluate the loss. It is no wonder that great attention is focused at the woodworks protection. For example, there are special laboratories in restoration workshops dealing with the problem.

When developing, common furniture beetle passes four stages: Ovum, larva, pupa, and adult insect the beetle (usually called imago). Common furniture beetle is 5 mm long, 1.2–1.7 mm wide, and umber colored. Their antennas are slightly shorter than hemi some, with three last elongated segments. Living insect has them stuck up to the front. The beetle becomes still when touched, and antennas are hidden in sternum hollow. The head is hidden under beetle pronotum and is almost invisible from above. The pronotum is hood shaped with clear callus in the middle. There are on the wing flaps 10 rows of equal and distinct dot striae on the wing flaps (Figure 1).

Wood fretter males are outwardly different from females. Females are bigger and, moreover, have the abdomen end relatively even, while males have a pronounced cross hollow.

In houses, some beetle individuals can be observed during the whole year, however, at the beginning of summer they are most abundant. This is time of the so called "flight", when females and males copulate. The name "flight" is not completely applicable to this pest species, because common furniture beetle only flies in a close range and very seldom, usually in warm summer days. The majority of beetles stay in places they have appeared or nearby. By this, low wood fretter inclination to flying the investigators explains frequent localization of damages caused by them. To this period of life, an original property of common furniture beetle to produce short ticking sound is confined. They are heard well in summer evenings, if there is wood infested by fretters in the house.

Being disturbed, the beetle draws down antennas and legs and becomes shocked, to say "feigns dead". This defense reaction helps the wood fretter to escape from enemies, because when drawing down legs it falls down and is then difficult to find.

The copulation usually proceeds in grooves and cracks abundant in dried wood and infrequently on its surface. Almost immediately after the copulation females begin to oviposit. The eggs are elongated, whitey, 0.5 mm long and 1.2 mm wide. They are only observed on wood with the help of a lens. Eggs are normally glued to the substrate and it is very difficult to detach them from it without damage. It is found that the around 80 eggs are ovipositor. After ovipositing, beetles usually die shortly. Their life span is from 6 to 22 days [2, 5, 6].

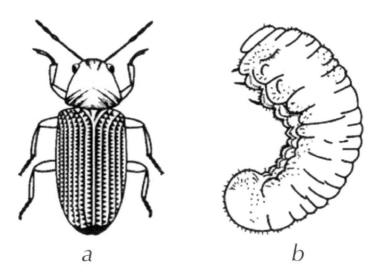

a   *b*

FIGURE 1   Wood fretter (a) and its larva (b)

The embryonic development (from fertilization till larva hatching) lasts 10–12 days. In this period, wood fretters are especially sensitive to the impact of external factors, namely to humidity and temperature. At 50% relative humidity, the egg survivability decreases, and at 45%, they die. Eggs are also sensitive to the impact of high temperature. If larvae die at +45°C, temperature of +30°C is already mortal for eggs.

After embryonic development, the larva gnaws through the end of the egg, which is pressed to wood, and penetrates into it.

Larvae of these beetles are C-shaped, they are whitey, have relatively big head, and slightly reddish vestiture from short thin hair. Making the wood interior rotten, they do not touch the external layer. This is the reason why infested wood be hardly distinguished from non-infested one, and only after rounded beetle flight holes have appeared the danger rate becomes clears.

It is noteworthy to mention that larvae of some wood fretters damage not only wood. They are able to live for the account of any phylogenic and sometimes zoogenic food. Some instances are known, when they developed in many generations consuming only opium or attacked dried meat, and so on.

The secret of high wood fretter adaptability to different nutrition was disclosed by studying their digestion features. It turned out that beetle larvae have a great variety of entreat enzymes, which help them to uptake not only sugars, proteins, and starch but also cellulose, the wood stable component. Moreover, there are special formations in their body (mycetomes), where specific species of microorganisms supplying the larva with the most unprocurable nitrogen-containing substance, exactly the substance contained in low amounts in wood multiply.

The role of symbionts in the wood fretter life is so great that they are passed on from generation to generation. When the female oviposits, the egg surface is covered by these microorganisms. When gnawing through the eggshell, young larva gets a portion of symbionts simultaneously, which then multiply in mycetomes. Thanks to these allies, wood fretter larvae can consume even cellulose. In the wild, however, they prefer an old, changed with time and by fungi dried wood.

Grown larvae are seemingly different from young larvae. They are bigger and, in contrast with young larvae, have spines on the back. By means of these spines, the larva leans against burrow walls when moving.

Before pupation, the larva moves to the wood surface leaving ungnawn layer of about 0.5 mm thick. Here the larva converts to pupa, which becomes the adult insect within 2–3 weeks, and the whole cycle repeats again.

All stages of common furniture beetle development from the egg to the adult insect lasts about 1 year, at least 9 months being spent at the larval stage. At this very stage, common furniture beetle causes most severe damages to woodworks and buildings.

The hatched young larva inserts itself into wood. The outlet hole is 0.1–0.2 mm in diameter and is only visible with the help of a lens. As the insect grows, the burrow width increases reaching 2.0–2.3 mm in diameter. Burrows are filled with residual digested wood, which is called wormhole dust. The feature of many wood fretters is that most of the burrows are concentrated in the springwood. Therefore, as wood is damaged severely, it is easily stratified into separate layers composed of autumn wood.

The specific feature of common furniture beetle is that it attacks woodworks, which have been in use for a definite period (5–25 years). This phenomenon is usually associated with the texture and physical state of the wood. The larvae of common furniture beetle can also normally develop on green wood.

Along with humidity, the main factor determining the damage rate of wooden buildings is the probability of eggs occurrence on them, which depends on multiplication and activity of fretter females. And as long as the last indices are low, a long time will pass before the building is oviposited.

The damage level of wood structures and woodworks depends on how conditions in apartment buildings are similar to the optimum of wood fretter development. The basic conditions are: temperature and humidity conditions of unheated building sections (garrets and basements). Temperature sufficient for development of wood fretters in garrets and basements can only be reached in summer, except of cases, when wood constructions under the leads are wetted due to defective roof. At that time, the relative air humidity is 60–80%.

Distinctly different conditions are formed in heated quarters of buildings. The basic indices of the microclimate in living quarters should be: temperature should be + (21–22)°C and humidity should be 30–45% in winter and in +(13–24)°C and 35–50% in summer, respectively.

At any time, deviations from above listed indices are observed. Nevertheless, temperature conditions in living quarters during the year correspond to the optimum for wood fretter development. The case is somewhat different for humidity. Common humidity in premises promotes quick wood drying. Wood humidity adjusts to air humidity and at normal temperature equals 11–17 abs. %. Therefore, air humidity in living quarters is quite unfavorable for of wood fretter eggs development, and wood humidity is lower, than necessitated for larvae. Common furniture beetle distribution foci are concentrated in places, where for any reasons air and wood humidity is increased.

The floor wood of ground floors, basements and kitchens, and the ends of wooden beams becomes wet most frequently. In wooden houses, the wood fretter is frequently met in lower timber sets of walls, beams, and the floor wood of ground floors. This pest is also detected in base moldings, especially at the external wall, and the bottom part of the door case. Inspections show that floors are first by the quantity of damages, followed by framing, doors, and door-cases. Partitions are damaged less frequently because of their relatively quick drying [2, 5, 6].

The appearance of common furniture fretters indicates either construction defects during building or improper operation of the building that increases air and structure humidity as compared with the norm.

If living beetles are found in a premise, the infested places in woodworks must be found. Only when boundaries of damage foci are determined, measures against wood fretters should be undertaken. The damage focus is usually determined by the presence of flight holes in wood. However, it should be determined, if this damage focuses is active that is if there are larvae in the wood, or all beetles have flown away. The presence of burrows indicates the end of the process. Precisely, this is the main difficulty. Common instructions on common furniture beetle control recommend paying attention to the occurrence of wormhole dust poured out from the burrows. It is assumed that this sign quite reliably testifies the presence of larvae in the wood. However, the X-ray method is more trustworthy, because it clearly shows larvae images on the film that absolutely guarantees that the active damage focus is found.

The X-ray method is normally used in museums to evaluate the woodwork infestation degree. However, in practice, they only are inspected. Meanwhile, the following signs should be taken into account: the presence of flight holes, the whole edge contamination rate (new holes have clean edges and fresh wood is visible), and wormhole dust pouring out from the burrows. However, the most reliable method for evaluation of the active focus is the disclosure of shy places. This should be made whenever possible.

In houses, a death watch beetle (*Anobium punctatum*) is also frequently observed. It is dark-brown, with 3–4 mm long cylindrical body, covered by thin gray indumentums. Its larvae damage furniture, frames, floor, ceiling beams, and wall beams. Furniture beetle (*A. Pertinax*) is slightly bigger and has two light spots in front angles of the pronotum. Its larvae usually appear on garret flooring, in angles of rooms, on floor boards, but they do not damage furniture.

Drugstore beetle (*Stegobium paniceum*) larvae are omnivorous rotting crackers, stale bread, furniture, dry insects, book covers, and other materials. In libraries, it is called the "book beetle"; in food warehouse, it is responsible for "wormy" crackers; in museums, its larvae damage stuffed animals. The beetle itself is plain, 2–3 mm long, and foxy colored. It inhabits living quarters, and in the evening, it flies toward light.

Control measures against common furniture beetle. All control measures are divided into three groups: (1) construction-economic; (2) chemical; and (3) physico-mechanical [2, 7, 8].

Prophylactic measures do not allow wood fretter damage of wooden structures, woodworks or furniture for a long time. They are construction economic and partly chemical methods. These measures should work as long as possible, because in the most cases they are only carried out during construction and repair; their application in operated buildings is very expensive and is connected with considerable difficulties.

Destructive are control measures used for extermination of wood fretters, already living in wood. They are physico-technical and partly chemical methods. Destructive measures completely exterminate wood fretter, however, for relatively short period of time. With respect to this division, the requirements to different control measures also change.

Constructive measures for wood protection against fretters represent the entire group of actions applied on the wood way from the cutting area to warehouses and living quarters. They include cutting practice, time for wood removal from the cutting area, storage regime, and wood use regulations. The main goal of constructive control measures is to limit the possibility of wood fretter appearance in wood or to stop their further development. In the built construction, wood fretters appear by two ways: during the summer or with infested wood.

Elimination of infesting necessitates refusal from storage and use of waste wood as firewood; old furniture shall only be used in flats after its thorough inspection for infestation by common furniture beetle. Of course, even if the measures are carried out thoroughly, the probability of wood fretter appearance in the flat remains. After finding favorable conditions and ovipositing, it will propagate and damage wood. Therefore, the second stage of control measures concludes arrangement of such conditions, under which eggs and larvae cannot develop.

Chemical control measures include the use of different poisons for fretter extermination. These poisons are called insecticides. By impact on fretters, insecticides are divided into three groups: intestinal, contact, and fumigants preparations affecting respiratory channels.

Intestinal insecticides affect wood fretters, when entered their intestine. Contact insecticides have an effect after physical contact with the beetle body surface. Fumigants poison the beetles after inhalation.

The main part of the life cycle (8–10 months) the wood fretter spends as larva and pupa inside the wood. The adult insect lives only 2–3 weeks after leaving the wood. This way of life significantly complicates the fretter control and put forward additional demands on the pest killer agents.

Insecticides applied against fretters must be highly toxic for insects and stored for a long time, when injected into the wood; they must be human-friendly, deteriorate physico-mechanical properties of the wood, and have sharply objectionable odor.

For the control of wood fretter larvae, the wood is processed by chemical agents, including hexachlorane, turpentine mixed with kerosene, wax, paraffin, and kreoline or a mixture of turpentine, kerosene and phenol. This means however do not ensure reliable and long-term wood protection against insects. Larvae are only completely exterminated after high-frequency current treatment of the infested wood.

Larva burrows and flight holes made by them during the tree growth, wood storage and use of furniture, parquet, plywood, timber, or chipboard are called wormholes.

The wormhole looks like grooves and oval holes of different depth and size. By depth, wormholes can be: surface up to 3 mm, shallow from 5 to 15 mm (in round timber), deep – more than 15 mm, and through. By size, wormholes are small not more than 3 mm and large more than 3 mm. These biological damages deteriorate the wood outlook; lower the strength and commercial wood output. Large quantity of wormholes reduces mechanical properties of the wood and increases waste. Only surface wormholes have impact on mechanical properties of the wood [10-12].

KEYWORDS

- **Destructive**
- **Eggs**
- **Larvae**
- **Pronotum**
- **Wood fretter**
- **Woodworks**

REFERENCES

1. Pekhtasheva, E. L. *Biodamages and protections of nonfood materials*. Moscow Masterstvo Publ. House, pp. 224 (in Russian) (2002).
2. *Actual problems of biological damage and protection of materials, components and structures*: *Sat. articles*. By N. A. Plate (Ed.), Nauka Publishers (Science): Scientific Council for biological damage of the USSR, Moscow, p. 256 (in Russian) (1989).
3. Proceeding of Conference. *Biological Damages*. By N. A. Plate (Ed.), Nauka Publishers (Science), Moscow, p. 226 (in Russian) (1978).
4. Problems of biological damage to materials. *Environmental aspects*. By N. A. Plate (Ed.), Scientific Council for biological damage of the USSR Publishing House, Moscow, p. 124 (1988).
5. Vorontsov, A. I. *Insect destroyers of wood*. By N. A. Plate (Ed.), Nauka Publishers (Science): Scientific Council for biological damage of the USSR, Moscow, p. 176 (1981).
6. Persians, M. P. *Furniture grinder and measures to combat it*. Leningrad, Nauka Publishers (Science), Moscow, p. 40 (1966).

7. Gumargalieva, K. Z. and Zaikov, G. E. *Biodegradation and biodeterioration of polymers. Kinetical aspects*, Nova Science Publ., New York, p. 210 (1998).
8. Semenov, S. A., Gumargalieva, K. Z., and Zaikov, G. E. *Biodegradation and durability of materials under the effect of microorganisms*, VSP International Science Publ., Utrecht, p. 199 (2003).
9. Polishchuk, A. Ya. and Zaikov, G. E. *Multicomponent transport in polymer systems*, Gordon & Breach, New York, p. 231 (1996).
10. Moiseev, Yu. V. and Zaikov, G. E. *Chemical resistance of polymers in reactive media*, Plenum Press, New York, p. 586 (1987).
11. Emanuel, N. M., Zaikov, G. E., and Maizus, Z. K. *Oxidation of organic compounds. Medium effects in radical reactions*, Pergamon Press, Oxford, p. 628 (1984).
12. Jimenez, A. and Zaikov, G. E. *Polymer analysis and degradation*, Nova Science Publ., New York, p. 287 (2000).

11 New use of Ionic Liquids

*Olesya N. Zabegaeva, Tat'yana V. Volkova,
Alexander S. Shaplov, Elena I. Lozinskaya,
Ol'ga V. Afonicheva, Mikhail I. Buzin,
Ol'ga V. Sinitsyna, and Yakov. S. Vygodskii*

CONTENTS

11.1 Introduction 121
11.2 Experimental.. 122
    11.2.1 Materials 122
    11.2.2 The Ils Synthesis.. 123
    11.2.3 Composites Preparation Technique 123
    11.2.4 Measurements 123
11.3 Results 123
    11.3.1 Anionic Polymerization of E-Caprolactam................... 123
    11.3.2 Influence of Ionic Liquids Nature on APCL................. 124
    11.3.3 Influence of The Il Concentration on The APCL............ 125
    11.3.4 Characterization of Composites..................................... 126
11.4 Conclusion... 134
Keywords 134
References... 134

11.1 INTRODUCTION

The aim of the suggested work was the investigation of different ionic liquids (ILs) in anionic ring-opening polymerization of ε-caprolactam (APCL) and studying structure and properties (tough, frictional and thermal, and etc.) of the obtained copolymers.

Polycaproamide (PCA) widely known as polyamide 6 is prepared by APCL and differed in outstanding mechanical and chemical properties [1]. This plastic is an important construction material. To increase the scale of its utilizations, it is necessary to tailor its properties by various additives.

Environmental friendly ILs has attracted the attention as an alternative to traditional toxic organic solvents and as new promising catalytic systems. Synthesis of new ILs with different nature of anions and cations stimulates the expansion of their use. For many reactions, solution of catalyst in the IL can be regarded as liquid reactor. The IL can be used not only as a reaction medium but also as polymeric materials plasticizers.

To date, there is only one way to obtain the composites based on PCA and IL, namely, filling by IL of previously prepared polymer melt [2, 3]. In anionic polymerization of lactams, it is necessary to enter additives at the step of the initial reagents preparation, because the formation of the polymer is combined with the process of forming products made of it (for example, rapid injection forming). Radical [4-6], cationic [7, 8], electrochemical polymerization [9, 10], and polycondensation [11, 12] in ILs have been investigated by many researchers. However, there are only a few papers devoted to the polymerization of (vinyl) monomers by anionic mechanism [13-15] in the presence of ILs. In this study, APCL in the presence of different ILs was studied. The results were compared with those obtained for APCL without ILs.

To study the effect of ILs (their structure and amount) on the properties of the PCA, several ILs (Table 1) were synthesized and examined in APCL process. All used ILs are not volatile and stable at the temperatures exceeding polymerization temperature as well as melting point of PCA and can be mixed with the molten monomer in any ratio.

TABLE 1   Ionic liquids used

Ionic liquids	Mp*, °C	Refs.
[1-Me-3-Etim]N(CF$_3$SO$_2$)$_2$	−3	16
[1-Me-3-Buim]N(CF$_3$SO$_2$)$_2$	−4	16
[1-Me-3-Buim]NO$_3$	2,06	17
[1-Me-3-Buim]PF$_6$	10	18
[1-Me-3-Buim]CF$_3$COO	(−50 ÷ −30)	16
[1-Me-3-Buim]CH$_3$COO	–	–
[1-Me-3-Buim]BF$_4$	(−81)	19
[1-Me-3-Buim]N(CN)$_2$	(−90)	20
[1-Me-3-Buim]Br	(−58)	17
[1-Me-3-C$_6$im] N(CF$_3$SO$_2$)$_2$	−7	21
[1-Me-3-C$_8$im] N(CF$_3$SO$_2$)$_2$	(−84)	21
[1-Me-3-C$_{14}$im] N(CF$_3$SO$_2$)$_2$	34,3	22
[3C$_6$PC$_{14}$] N(CF$_3$SO$_2$)$_2$	(−76)	23
[N-Me-N-BuPyrr] N(CF$_3$SO$_2$)$_2$	−15	24

* Glass transition temperature is given in brackets.

11.2   EXPERIMENTAL

11.2.1   Materials

Monomer, namely ε-caprolactam (CL), was recrystallized from benzene and dried thoroughly in high vacuum at 50°C before use [25]. The catalyst, CL magnesium bromide, was prepared by the interaction of synthesized ethyl magnesium bromide with CL, according to the published procedure [26]. The obtained catalyst structure estimated by IR spectroscopy method is fully coincided with data. [26].

11.2.2   The ILs Synthesis

All used ILs were synthesized in accordance with the published elsewhere [27-31]. The synthesized ILs were characterized by elemental analysis, NMR, and IR spectroscopy prior to use. The purity of ILs estimated by these methods was higher than 98%.

11.2.3 Composites Preparation Technique

All the manipulations were carried out under inert atmosphere. The APCL in the presence of ILs was performed in the glass tubes equipped with an argon inlet and outlet. After CL melting of at 150°C, the calculated amount of IL was added. Then the separately prepared solution of CL magnesium bromide in CL was quickly added. The reaction mixture was stirred for 1 min, and finally the reaction mixture was kept at 150°C for 1.5 hr [25]. The reaction parameters as well as the inherent viscosity of the obtained polymers are summarized in the Tables 2 and 3.

11.2.4 Measurements

The IR spectra were recorded using Nicolet Magna-750 Fourier IR spectrometer. Inherent viscosities (η_{inh}) were measured using Ostwald capillary viscometer (0.05 g of polymer in 10.0 ml of HCOOH at 25.0°C).

The notched Izod impact strength and flexural modulus for polymer blocks were determined on PDM-10 and "Dinstat" devices, respectively [32]. Thermofrictional examinations were measured using I-47 butt friction machine [32].

The glass transition temperatures (T_g) of the composites were determined by differential scanning calorimetry method (DSC) using a Mettler-822e instrument at heating rate of 20 deg/min and in an inert gas atmosphere.

The composite blocks surface (the chipped) was scanned using Supra 50VP high-resolution scanning electron microscope (Japan). Samples were prepared by cleaving after the freezing of the polymer blocks in liquid nitrogen. Thermogravimetric analysis (TGA) was performed in air with a MOM-Q1500 derivatograph instrument (Hungary) at heating rate 5 deg/min.

11.3   RESULTS

11.3.1   Anionic Polymerization of ε-caprolactam

The APCL was carried out under the action of 0.35 mol % of catalytic system consisting of CL magnesium bromide and N-acetyl-ε-caprolactam (1:1) in the presence of above-mentioned ILs (from 2 to 10 mol %) and at constant temperature equal to 150°C. It is well known [33, 34] that polymerization of CL and the resulting polyamide crystallization, depending on conditions, occur sequentially or simultaneously. The rate of polymerization was estimated by the time of the loss of mobility of the reaction system. The validity of such characteristic has been shown using the method of isothermal calorimetry on APCL process performed with model compounds, namely, with diimides [25, 35, 36]. Based on the earlier results [33], the onset of the PCA crystallization is determined by time of the reaction system turbidity. The completion crystallization was assessed by time of the polymer block detachment from the walls of the reactor vessel.

11.3.2   Influence of Ionic Liquids Nature on APCL

Table 2 summarizes the results of the APCL in the presence of ILs (2 mol %). Apparently , the nature of the anion of IL has a more pronounced effect on the rate of CL polymerization than that of the nature of its cation. For example, the polymerization times of CL without IL and in the presence of [1-Me-3-Etim]N(CF$_3$SO$_2$)$_2$, [3C$_6$PC$_{14}$] N(CF$_3$SO$_2$)$_2$, or [N-Me-N-BuPyrr]N(CF$_3$SO$_2$)$_2$ were equal (Table 2, Entries 1, 2, 14, 15). However, with an increase in the number of C atoms of alkyl substituent of the cation IL from 2 to 14, the polymerization time increases monotonically (Table 2, Entries 2, 5, 11–13). This phenomenon can be explained by the steric factors. Because of the increasing number of carbon atoms in the imidazolium cycle alkyl substituent, the dynamic viscosity of IL increases and therefore, the viscosity of reaction system raises [37-39].

TABLE 2   Anionic polymerization of CL in the presence of IL*.

№	Ionic liquid	Time[1], min			P[2], %	η_{inh}, dl/g
		A	B	C		
1	–	3	5	7	2	2.3
2	[1-Me-3-Etim]N(CF$_3$SO$_2$)$_2$	3	4	9	10	2.2
3	[1-Me-3-Buim]PF$_6$	2.5	4	27	23	2.3
4	[1-Me-3-Buim]NO$_3$	34	50	–	10	2.3
5	[1-Me-3-Buim]N(CF$_3$SO$_2$)$_2$	5	6	30	10	2.4
6	[1-Me-3-Buim]CF$_3$COO	38	45	–	46	1.1
7	[1-Me-3-Buim]CH$_3$COO	–	–	–	–	–
8	[1-Me-3-Buim]BF$_4$	–	–	–	–	–
9	[1-Me-3-Buim]N(CN)$_2$	–	–	–	–	–
10	[1-Me-3-Buim]Br	–	–	–	–	–
11	[1-Me-3-C$_6$im] N(CF$_3$SO$_2$)$_2$	8	8	30	11	2.4
12	[1-Me-3-C$_8$im] N(CF$_3$SO$_2$)$_2$	10	7	–	12	2.5
13	[1-Me-3-C$_{14}$im] N(CF$_3$SO$_2$)$_2$	17	15	–	13	2.4
14	[3C$_6$PC$_{14}$] N(CF$_3$SO$_2$)$_2$	3	3	8	17	2.3
15	[N-Me-N-BuPyrr]N(CF$_3$SO$_2$)$_2$	4	5	11	20	2.3

* IL concentration is 2 mol %
[1]Time of loss of mobility (A), turbidity of the reaction system (B), and the polymer block detachment from the walls of the reactor vessel (C)
[2]P -portion soluble in methanol

Influence of the anion nature on the polymerization rate is illustrated by using 1-methyl-3-buthylimidazolium ILs. The rate of APCL grows up upon adding [1-Me-3-Buim]PF$_6$ (entries 3). On the contrary, polymerization rate in the presence of IL bearing NO$_3^-$ and CF$_3$COO$^-$ anions (entries 4, 6) was very low and the APCL does not occur at all when ILs contain BF$_4^-$, CH$_3$COO$^-$, N(CN)$_2^-$ and Br$^-$ anions (entries 7–10). It is possible that the polymerization rate reduction or its complete reaction cessation in the presence of cited salts is due to the poisoning the catalyst system. These results are consistent with the ring-opening polymerization of CL in ILs [40, 41]. The authors [41] believe that the strong interaction between catalyst and the anion of ILs is the reason of CL polymerization acceleration or retardation.

Despite slowdown in the rate of polymerization in some cases, the crystallization of PCA in the presence of IL (ILs with NO$_3^-$ and CF$_3$COO anions are the exceptions) began earlier than in the absence of IL. The fact of the simultaneous polymerization and crystallization in the synthesis of polyamides by anionic polymerization of lactams is reported in several papers [42, 43]. Although, the crystallization of PCA in the presence of IL occurs simultaneously with the polymerization or immediately after its final step, in most cases, its full completion is necessary twice or more time than that of the homopolymerization of CL. The reason for such behavior is, probably, the dramatic increase of the reaction system viscosity, which prevents PCA crystallization.

Inherent viscosities for all resultant composites (exception [1-Me-3-Buim] CF$_3$COO$^-$) are similar to PCA obtained without ILs.

11.3.3   Influence of the IL Concentration on the APCL

Influence of IL concentration on APCL was studied by adding to a reaction system of 2, 5, 7, and 10 mol % [1-Me-3-Buim]N(CF$_3$SO$_2$)$_2$. Results are presented in the Table 3. As can be seen, the rate of polymerization slows down with increased concentrations of IL. Upon the introduction of 10 mol % of IL, the polymerization does not proceed at all. This can be explained by rapid increase in the viscosity of the reaction system. Initial reagents and reaction products can greatly affect the viscosity of the IL during chemical reactions in a medium of the IL. The increase in the IL viscosity itself leads to an increase in viscosity of the reaction system. This is due to the fact that the ILs associates can easily interact with the diluents, especially polar ones. For example, due to the presence of halogens, the viscosity of IL increases dramatically. This phenomenon can be explained by the formation of hydrogen bonds between the halogen ion and imidazole [44]. In this case, it is bromide ions, which are present in the system due to the catalyst dissociation.

It should be noted that some features of the CL polymerization are present at the change of IL concentration in the reaction system. At 2 mol % of IL, the APCL occurs homogeneously and its rate is similar to the rate of CL polymerization without IL. With increase in concentration of IL to 5 mol %, the white suspended matter is appeared in the reaction mixture and the rate of polymerization slows down (Table 3, Entry 3). Upon adding 7 mol % IL, the amount of the white suspended matter significantly is increased; the time of polymerization rises significantly (Table 3, Entry 4). Sekiguchi [43] observed a similar phenomenon when the rate of polymerization slows down at the appearance of a white suspended matter in the reaction system. Apparently, in these cases, CL polymerization and the resulting polymer crystallization occur si-multaneously [45]. The observed decrease in the rate of CL polymerization in the pres-ence of IL may be also due to the fact that the early onset of polymers crystallization leads to the immurement of the active centers, which located at the ends of growing chains, thereby preventing access lactam anions [43].

Nevertheless, inherent viscosities for all resulting composites are high enough (about 2.3 dl/g) and similar to PCA obtained without ILs.

TABLE 3   Anionic polymerization of CL in the presence of 1-methyl-3-buthylimidazolium b is (trifluoromethylsulfonyl) amide.

№	[IL], mol.%	Time[1], min			F[2],%	η_{inh}, dl/g
		A	B	C		
1	–	3	5	7	2	2.3
2	2	5	6	30	10	2.3
3	5	9	9	30	20	2.4
4	7	19	17	35	65	2.4

[1]Time of loss of mobility (A), turbidity of the reaction system (B), and the polymer block detachment from the walls of the reactor vessel (C)
[2]P- portions soluble in methanol

11.3.4   Characterization of Composites

All the obtained copolymers are either white or slightly yellowish substances. The polymer sample surface is not oily despite ILs' occurrence. The filled polymers are partially soluble in methanol exceeding those for the homo-PCA (Tables 2 and 3). Such fact can be explained by the complete wash out of IL from the polymer sample during the methanol extraction. Moreover, the IR spectra of the PCA obtained in the presence of 2 mol % [1-Me-3-Buim]N(CF$_3$SO$_2$)$_2$ before extraction (Figure 1(b, 2)) contain bands characteristic for IL. Thus, the signals at 1,350–1,000 cm^{-1} and 617 cm^{-1} relate to the bis(trifluoromethylsulfonyl)amide anion. After extraction of this sample with methanol, bands characteristic of IL disappears.

We believe that IL can certainly influence on the growth and structure of supra-molecular entities that affect on the final product properties because, it had a strong influence on the APCL process. The microstructure of composite blocks containing ILs was studied on the chipped samples using electron scanning microscopy. From the images (Figure 2), a significant influence of IL amount on the nature of the supra-molecular structure of the polymer composites can be seen. This influence is initially expressed in the gradual decrease in the size of structural elements of the blocks that contain 1–2 mol % IL, and then to their enlarge with increase of IL concentration. The structure of the unfilled PCA consists of spherulites. The spherulites consist of radially symmetric fibers. The diameter of spherulites is 35–45 μm (Table 4).

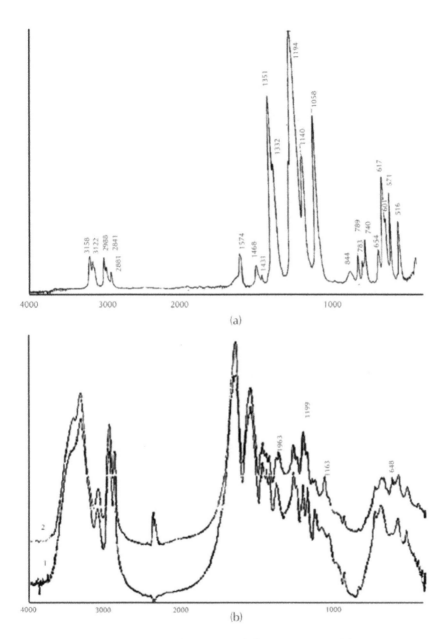

FIGURE 1 The FTIR spectra of [1-Me-3-Buim]N(CF$_3$SO$_2$)$_2$ (a),plain PCA (b,1) and its composite with [1-Me-3-Buim]N(CF$_3$SO$_2$)$_2$ (b,2).

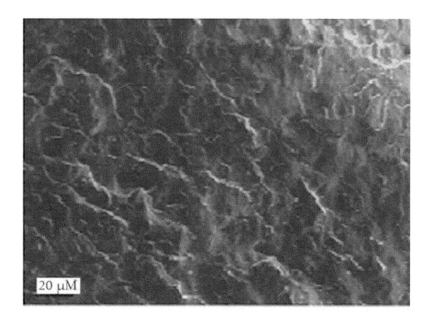

a

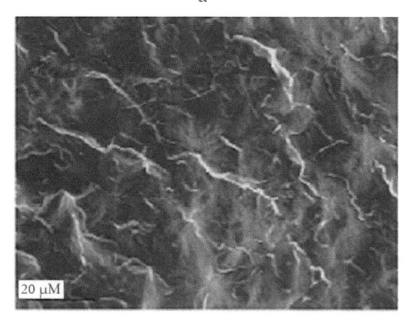

b

FIGURE 2   *(Continued)*

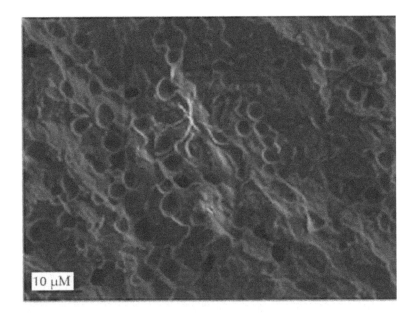

c

d

FIGURE 2   The SEM image of PCA blocks surface (a) and composite blocks containing of IL mol %: 1(b), 2(c), and 7(d).

TABLE 4   Microstructure of composite blocks containing ILs.

[IL], mol %	Diameter of spherulites, μm	α*, %
0	45	58 (78)
1	30–40	58 (65)
2	20	62
7	30–33	56

*α – Degree of crystallinity (in the brackets for samples extracted with methanol)

The spherulites are densely packed and have a clear boundary. In sample containing 1 mol % IL, the sizes of spherulites are 30–40 μm. In the SEM images of PCA, containing 2 mol % IL, an inhomogeneous porous structure is observed (Figure 2(c)). Nevertheless, the individual spherulites with size less than 20 μm can be distinguished. A further increase of the IL concentration in the polymer leads to an enlargement of the structural elements. Sheaves shape spherulites of 30–32 μm in diameter were detected on the surface of composite blocks containing 7 mol % IL (Figure 2(d)). The amplification of spherulite aggregation increases with IL concentration rise. Figure 2(d) shows that the spherulites interact with one another upon the formation of agglomerates. This interaction is accompanied by deformation of the spherulites and leads possibly to the interpenetration of composite components. There are some pores, large, and small voids between agglomerates. Apparently, the spaces between spherulites are filled with IL. Consider the reasons of changes of the nature of supramolecular structures and the results they demonstrate, it can be said that IL can play the role of nucleation and affect on the forming macromolecules crystallization and their growth. Upon increase in IL concentration in the reaction mixture, its particles begin to coalesce, forming large aggregates. Thus, there is a decrease of the total number of crystallization centers and consequently, increase of the size of spherulites.

The crystalline structure of customary PCA and PCA-based copolymers obtained in this work was investigated by the DSC method. Table 4 demonstrates the influence of IL on the relevant PCA composites crystallinity. The presence of residual monomer in the polymer decreases PCA crystallinity degree on ~25%. It was assumed that the presence of IL and especially an increase in its concentration will further reduce this polyamide structural parameter. Nevertheless, the degree of crystallinity of the filled PCA is around 60%, and is slightly differed from the crystallinity of unmodified one (58%).

The T_g of PCA and its composites with ILs were determined by the DSC measurements. Generally, crystalline polymers are characterized by a slight change of heat capacity in the glass transition temperature region. Therefore, the synthesized samples were quenched in liquid nitrogen. This procedure allowed to increase the heat capacity change at the glass transition temperature and make the T_g determination more accurate (Figure 3). The T_g of PCA obtained with 2 mol % of IL was more than 10 degrees below the corresponding parameter for standard PCA and do not depend on the nature of IL (Table 5). The T_g decreases monotonically with an increase in concentration of IL.

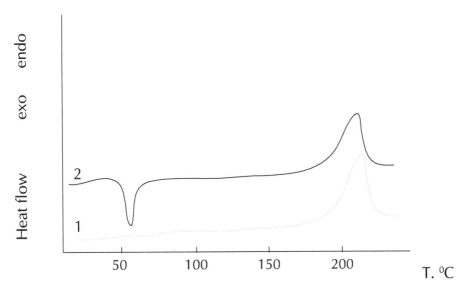

FIGURE 3 The DSC curves of PCA before (1) and after (2) quenching in liquid nitrogen. Heating rate is 20°C/min.

The application of PCA at high temperatures is limited by its comparatively low thermal stability. It was assumed that thermal stability of the PCA increases due to the presence of heat resistance IL. Thermal stability of modified PCA was estimated by the temperature loss of the sample of 5% wt ($T_{5\%}$) and was found to depend on several factors. In such a way, the nature of anion plays an important role. Polyamides, obtained in the presence of 2 mol % IL, are characterized by increased $T_{5\%}$, equal about 300°C in contrast with unmodified PCA (Table 5).

TABLE 5 The properties of the copolymers consisted from PCA and 2 mol % of IL.

Ionic liquid	T_g, °C	T_{mp}, °C	$T_{5\%}$, °C	$T_{10\%}$, °C
–	48 (56)[1]	220 (217)[1]	280 (340)[1]	350
[1-Me-3-Etim]N(CF$_3$SO$_2$)$_2$	25 (38)[1]	220 (215)[1]	290	330
[1-Me-3-Buim]N(CF$_3$SO$_2$)$_2$	26	218	305	320
[1-Me-3-C$_6$im] N(CF$_3$SO$_2$)$_2$	27	215	305	320
[1-Me-3-C$_8$im] N(CF$_3$SO$_2$)$_2$	29	219	305	320
[1-Me-3-C$_{14}$im] N(CF$_3$SO$_2$)$_2$	32	217	305	320

TABLE 5   *(Continued)*

Ionic liquid	T_g, °C	T_{mp}, °C	$T_{5\%}$, °C	$T_{10\%}$, °C
[1-Me-3-Buim]N(CF$_3$SO$_2$)$_2$[2]	14	213	305	315
[1-Me-3-Buim]PF$_6$	28	217	295	340
[1-Me-3-Buim]NO$_3$	26	216	300	350
[1-Me-3-Buim]CF$_3$COO	26	216	310	365

[1]Samples after methanol extraction
[2]Sample obtained in the presence 5 mol % of IL

Nevertheless, the further heating above 300°C leads to significant chemical changes in the samples synthesized with IL. Thus, at the temperature of 350°C the samples of homo-PCA and composites obtained in the presence of 1-methyl-3-butylimidazolium IL with PF$_6^-$, NO$_3^-$, and CF$_3$COO$^-$ anions lose no more than 10% by mass, while the mass of the composites containing 1-methyl-3-butylimidazolium IL with bis(trifluoromethylsulfonyl)amide anion at this temperature is decreased by 18–20%. This behavior can be explained by the substantial difference in compatibility of PCA with selected ILs. It was found [46] that the treatment of polyamides in ILs at high temperatures (above 350°C) results in depolymerization to give monomers and their decomposed compounds. The rate of depolymerization is largely dependent on solvent polarity. Upon greater polarity of IL, the rate of PCA depolymerization becomes higher. The presence of oxygen containing groups in the anion reduces the polarity of IL due to very strong hydrogen bonds between the imidazolium cation and anion [47]. The PF$_6$ anion also can form strong hydrogen bonds with imidazolium anion. 1-Methyl-3-butylimidazolium bis(trifluoromethylsulfonyl)amide IL has the highest polarity in a series of considered ILs. Therefore, composite PCA, resulting in the presence of [1-Me-3-Buim]N(CF$_3$SO$_2$)$_2$, has the lowest thermal stability.

The mechanical properties of the composites were studied in detail and are summarized in Table 6. All composites PCA obtained in the presence of 2 mol % IL have practically the same flexural modulus (~150 MPa). At the same time, the investigated samples demonstrated the behavior similar to viscoelastic materials. In such a way, composites obtained with [N-Me-N-BuPyrr]N(CF$_3$SO$_2$)$_2$ bend without demolition up to 90° corner, maintaining a pressure of 150 MPa. For comparison, unmodified PCA bends to the corner no higher than 40° and breaks at the stress of 150 MPa. All novel composites obtained with 2 mol % of [1-Me-3-Etim] and [1-Me-3-Etim] IL regardless of the anion type break as well as PCA at the pressure of 150 MPa and bend up to ~70° corner. At the same time, increasing the size of the alkyl radical imidazolium IL as well as the use of the [3C$_6$PC$_{14}$] N(CF$_3$SO$_2$)$_2$ leads to an increase of the flexural modulus and the resulting samples bend without the fracture up to 60–70° corner. The introduction of IL concentration in the polymer matrix leads to a significant flexural strength reduction (Table 6, Entries 7, 8). It was found what composites obtained with ILs in contrast of PCA demonstrated more than double decrease in flexibility modulus. Flexibility modulus is significantly decreased with the increase in IL concentration. All composites obtained in the presence of IL (2 mol %) are not destroyed upon the notched Izod impact strength tests in contrast to plain PCA. Composites synthesized with 5 or 7 mol % of IL showed impact strength decrease (Table 6, Entries 7, 8).

TABLE 6   Mechanical properties of the copolymers composed from PCA and 2 mol % of IL.

№	Ionic liquid	Flexural strength[1], MPa	Compression		Notched Izod impact strength[1], Aj, kJ/m²
			σ, MPa	E*10⁻³, MPa	
1	–	150/b (43)	52	1.7	5.4/b
2	[1-Me-3-Etim]N(CF₃SO₂)₂	150/b (53)	24	0.8	7.1/nb
3	[1-Me-3-Buim]N(CF₃SO₂)₂	150/b (68)	24	0.8	8.9/nb
4	[1-Me-3-C₆im] N(CF₃SO₂)₂	150/nb (67)	24	0.8	8.9/nb
5	[1-Me-3-C₈im] N(CF₃SO₂)₂	150/nb (68)	24	0.7	9.1/nb
6	[1-Me-3-C₁₄im] N(CF₃SO₂)₂	150/nb (68)	24	0.8	9.2/nb
7	[1-Me-3-Buim]N(CF₃SO₂)₂²	110/b (57)	19	0.7	5.5/b
8	[1-Me-3-Buim]N(CF₃SO₂)₂³	60/b (57)	15	0.6	5.2/b
9	[1-Me-3-Buim]PF₆	150/b	23	0.8	8.8/nb
10	[1-Me-3-Buim]NO₃	150/b (60)	22	0.8	6.1/nb
11	[1-Me-3-Buim]CF₃COO	150/b	26	0.8	5.8/nb
12	[3C₆PC₁₄] N(CF₃SO₂)₂	150/nb (57)	42	0.7	5.3/nb
13	[N-Me-N-BuPyrr]N(CF₃SO₂)₂	150/nb (90)	24	0.8	10/nb

[1]Sample breaks (b) and do not break (nb) at the stress. Corner at the bend is in brackets.
[2]Sample obtained with 5 mol % of IL.
[3]Sample obtained with 7 mol % of IL.

It is well known that PCA is widely used in friction devices. It was noted [48, 49] that ILs cause an improvement of the tribological properties of the PCA and polystyrene. The cited works in contrast with our method of approach of obtaining composites by blending a melt of synthesized polymers is described. Therefore, the investigation of copolymers tribological properties is of particular interest. Table 7 demonstrates the influence of IL on the friction behavior of the relevant PCA composites. As it is seen from this table, composites with IL are different in reduced temperature of frictional contact, decreased deterioration of the polymeric sample, and downturn of friction coefficient in comparison with standard PCA. It was established that all the studied polymers demonstrate stable friction without any mass transfer from polymer sample to the steel rider. In contrast to that, the friction of unmodified PCA was accompanied by a scratch intensifying in time, as well as by the transportation of polymer traces on a contact body.

TABLE 7   Frictional Properties of the copolymers composed of PCA and 2 mol % of IL.

Ionic liquid	Temperature of frictional contact, °C		Friction coefficient		Deterioration of the polymeric sample, %
	1 hr	2 hr	1 hr	2 hr	
–	57	70	0.69	0.94	100*
[1-Me-3-Etim]N(CF$_3$SO$_2$)$_2$	33	44	0.4	0.43	60
[1-Me-3-Buim]N(CF$_3$SO$_2$)$_2$	43	45	0.6	0.75	60
[N-Me-N-BuPyrr]N(CF$_3$SO$_2$)$_2$	43	53	0.25	0.6	60

* Deterioration of the polymeric sample is 8×10^{-4} g.

11.4   CONCLUSION

Summarizing the obtained results, it can be concluded that IL chemical structure greatly influences the yield and the properties of nascent composites.

It was established that application of IL in APCL allows to modify PCA and gain the desirable control over the phase composition, compression modulus, notched Izod impact strength, temperature of frictional contact, friction coefficient, and the other polymer characteristics.

KEYWORDS

- **Anionic ring-opening polymerization of ε-caprolactam**
- **Differential scanning calorimetry method**
- **ε-Caprolactam**
- **Ionic liquids**
- **Polycaproamide**
- **Thermogravimetric analysis**

REFERENCES

1. Puffr, R., Kuba′nek, V. In *Lactam-Based Polyamides*. J. Šebenda, R. Puffr, M. Raab, and B. Doležel (Eds.), Properties, CRC Press, Boca Raton, Vol.1. p. 29, p.187 (1991).
2. Sanes, J., Carri′on, F. J., Jim′enez, A. E., and Berm′udez, M. D. Influence of temperature on PA 6–steel contacts in the presence of an ionic liquid lubricant. *Wear.*, **263**, 658 (2007).
3. Sanes, J., Carrio, F. J., Bermu, M. D., and Martinez-Nicola, G. S. Ionic liquids as lubricants of polystyrene and polyamide 6-steel contacts. Preparation and properties of new polymer-ionic liquid dispersions. *Tribology Letters.*, **21**(2), 121 (2006).
4. Hong, K., Zhang, H., Mays, J. W., Visser, A. E., Brazel, C. S., Holbrey, J. D., Reichert, W. M., and Rogers, R. D. Conventional free radical polymerization in room temperature ionic liquids: A green approach to commodity polymers with practical advantages. *Chem. Commun.*, **13**, 1368 (2002).
5. Ma, H., Wan, X., Chen, X., and Zhou, Q. F. Revers atom transfer radical polymerization of methyl methacrylate in imidazolium ionic liquids. *Polymer.*, 44, 5311 (2003).
6. Strehmel, V., Laschewsky, A., Wetzel, H., and Goernitz, E. Free radical polymerization of n-butyl methacrylate in ionic liquids. *Macromolecules*, **39**, 923 (2006).

7. Vijayaraghavan, R., MacFarlane D. R. Living cationic polymerization of styrene in an ionic liquid. *Chem. Commun.*, **6**, 700 (2004).
8. Vijayaraghavan, R. and MacFarlane, D. R. Organoborate acids as initiators for cationic polymerization of styrene in an ionic liquid medium. *Macromolecules*, **40**, 6515 (2007).
9. Sekiguchi, K., Atobe, M., and Fuchigami, T. Electropolymerization of pyrrole in 1-ethyl-3-methylimidazolium trifluoromethanesulfonate room temperature ionic liquid. *Electrochem. Commun.*, **4**, 881 (2002).
10. Murray, P. S., Ralph, S. F., Too, C. O., and Walace, G. G. Electrosynthesis of novel photochemically active inherently conducting polymers using an ionic liquid electrolyte. *Electrochim. Acta.*, **51**, 2471 (2006).
11. Vygodskii, Ya. S., Lozinskaya, E. I., Shaplov, A. S., Lyssenko, K. A., Antipin, M. Y., and Urman, Y. G. Implementation of ionic liquids as activating media for polycondensation processes. *Polymer.*, **45**, 5031 (2004).
12. Lozinskaya, E. I., Shaplov, A. S., and Vygodskii, Y. S. Direct polycondensation in ionic liquids. *Eur. Polym. J.*, **40**, 2065 (2004).
13. Biedron, T. and Kubisa, P. Imidazolium ionic liquids with short polyoxyethylene chains. *J. Polym. Sci. Part A; Polym. Chem.*, **46**, 6961 (2008).
14. Kokubo, H. and Watanabe, M. Anionic polymerization of methyl methacrylate in an ionic liquid. *Polym Advan. Technol.*, **19**, 1441 (2008).
15. Vijayaraghavan, R., Pringle, J. M., and MacFarlane, D. R. Anionic polymerization of styrene in ionic liquids. *Eur. Polym. J.*, **44**, 1758 (2008).
16. Bonhote, P., Dias, A. P., Papageorgiou, N., Kalyanasundaram, K., and Gratzel, M. Hydrophobic, Highly Conductive Ambient-Temperature Molten Salts. *Inorg. Chem.*, **35**(5), 1168 (1996).
17. Gruzdev, M. S., Ramenskaya, L. M., Chervonova, U. V., and Kumeev, R. S. Preparation of 1-butyl-3-methylimidazolium salts and study of their phase behavior and intramolecular intractions. *Russian J. General Chemistry.*,**79**(8), 1720 (2009).
18. Jonathan, G. H., Visser, A. E., Reichert, M. W., Heather, D. W., Grant, A., and Rogers, R. D. Characterization and comparison of hydrophilic and hydrophobic room temperature ionic liquids incorporating the imidazolium cation. *Green Chem.*, **3**, 156 (2001).
19. Huddleston, J. G., Willauer, H. D., Swatloski, R. P., Visser, A. E., and Rogers, R. D. Room temperature ionic liquids as novel media for 'clean' liquid-liquid extraction. *Chem. Commun.*, (16), 1765 (1998).
20. Fredlake, C. P., Crosthwaite, J. M., Hert, D. G., Aki, S. N., and Brennecke, J. F. Thermophysical Properties of Imidazolium-Based Ionic Liquids. *J. Chem. Eng.*, **49**(4), 954 (2004).
21. Crosthwaite, J. M., Muldoon, M. J., Dixon, J. N. K., Anderson, J. L., and Brennecke, J. F. Phase transition and decomposition temperatures, heat capacities and viscosities of pyridinium ionic liquids. *J. Chemical Thermodynamics*, **37**(6), 559 (2005).
22. Bradley, A. E., Hardacre, C., Holbrey, J. D., Johnston, S., and McMath, S. E. J., Nieuwenhuyzen M. Small-Angle X-ray Scattering Studies of Liquid Crystalline 1-Alkyl-3-methylimidazolium. *Salts Chem. Mater.*, **14**(2), 629 (2002).
23. Rico, E. S., Corley, C., Robertson, A., and Wilkes, J. S. Tetraalkylphosphonium-based ionic liquids. *J. Organometallic Chemistry.*, **690**(10), 2536 (2005).
24. Tokuda, H., Ishii, K., Susan, Md. A. B. H., Tsuzuki, S., Hayamizu, K., and Watanabe, M. Physicochemical Properties and Structures of Room-Temperature Ionic Liquids. 3. Variation of Cationic Structures. *J. Phys. Chem. B.*, **110**(6), 2833 (2006).
25. Vygodskii, Ya. S., Volkova, T. V., Pashkova, O. N., Batalova, T. L., Dubovik, I. I., and Chekulaeva, L. A., Garbuzova I.A. Anionic polymerization of ε-caprolactam and its copolymerization with ω-dodecanelactam in the presence of aromatic polyimides. *Polymer Sci. J.*, **48**, 885; *Chem Abstr.*, **146**, 45768 (2006, 2007).
26. Zaharkin, L. I., Frunze, T. M., Gavrilenko, V. V., Kurashev, V. V., Chekulaeva, L. A., Kotel'nikov, V. A., Danilevskaya, L. B., Markov, A. V., Ur'ev, V. P., Boyarkin, M. A., Gebarov, O. G., and Egorov, A. M. USSR Pat. 1,641,824, 1991. *Chem Abstr.*, **116**, 74921m (1992).

27. Bonhôte, P., Dias, A. P., Papageorgiou, N., Kalyanasundaram, K., and Grätzel, M. Hydrophobic, Highly Conductive Ambient-Temperature Molten Salts. *Inorg. Chem.*, **35**(5), 1168 (1996).
28. Wilkes, J. S. and Zaworotko, M. J. Air and water stable 1-ethyl-3-methylimidazolium based ionic liquids. *J. Chem. Soc. Chem. Commun.*, 965 (1992).
29. Harlow, K. J., Hill, A. F., and Welton, T. Convenient and General Synthesis of Symmetrical N,N'-Disubstituted Imidazolium Halides. *Synthesis.*, (6), 697 (1996).
30. Cieniecka-Rosonkiewicz, A., Pernak, J., Kubis-Feder, J., Ramani, A., Robertson, A. J., and Seddon, K. R. Synthesis, anti-microbial activities and anti-electrostatic properties of phosphonium-based ionic liquids. *Green Chem.*, **7**, 855 (2005).
31. MacFarlane, D. R., Meakin, P., Sun, J., Amini, N., and Forsyth, M. Pyrrolidinium Imides: A New Family of Molten Salts and Conductive Plastic Crystal Phases. *J. Phys. Chem. B.*, **103**(20), 4164 (1999).
32. Vygodskii, Ya. S., Krasnov, A. P., Fedorova, L. S., Saharova, A. A., Afonicheva, O. V., and Volkov, I. O. *Polymer Sci. J.*, **41**, 74 (1999).
33. Wittmer, P. and Gerrens, H. Über die anionische schnellpolymerisation von caprolactam, *Makromol. Chem. B.*, **89**(1), 27 (1965).
34. Sekiguchi, H. and Coutin, B. Polymerization ability and related problems in the anionic polymerization of lactams. *J. Polym. Sci. A 1.*, **11**(7), 1601 (1973).
35. Vygodskii, Ya. S., Volkova, T. V., Batalova, T. L., Sapozhnikov, D. A., Dubovik, I. I., and Chekulaeva, L. A. Anionic polymerization of ε-caprolactam in the presence of aromatic poly(imides) as macromolecular activators. *Polymer Science J. A.*, **45**, 85; *Chem Abstr* 2003, **139**, 53412.3 (2003).
36. Vygodskii, Ya. S., Volkova, T. V., Batalova, T. L., Sapozhnikov, D. A., Nikiforova, G. G., Chekulaeva, L. A., Lomonosov, A. M., and Filatova, A. G. Anionic polymerization of ε-caprolactam in the presence of aromatic diimides. *Polymer Science J. A.*, **47**, 1077; *Chem. Abstr.*, **144**, 171400.4 (2005, 2006).
37. Hapiot, P. and Lagrost, C. Electrochemical reactivity in room-temperature ionic liquids. *Chem. Rev.*, **108**, 2238 (2008).
38. Crosthwaite, J. M., Muldoon, M. J., Dixon, J. N. K., Anderson, J. L., and Brennecke, J. F. Phase transition and decomposition temperatures, heat capacities and viscosities of pyridinium ionic liquids. *J. Chemical Thermodynamics.*, **37**(6), 559 (2005).
39. McLean, A. J., Muldoon, M. J., Gordon, C. M., and Dunkin, I. R. Bimolecular rate constants for diffusion in ionic liquids. *Chem. Comm.*, 1880 (2002).
40. Nomura, N., Taira, A., Nakase, A., Tomiokay, T., and Okada, M. Ring-opening polymerization of lactones by rare-earth metal triflates and by their reusable system in ionic liquids. *Tetrahedron.*, **63**, 8478 (2007).
41. Oshimura, M., Takasu, A., and Nagata, K. Controlled Ring-Opening Polymerization of ε-Caprolactone Using Polymer-Supported Scandium Trifluoromethanesulfonate in Organic Solvent and Ionic Liquids. *Macromolecules,*. **42**(8), 3086 (2009).
42. Kuskova, M., Roda, J., and Kralicek, J. Polumerization of lactams, 22. Effect of reaction conditions on the anionic polymerization of 2-pirrolidone (Part II). *Macromol. Chem.*, **179**(2), 337 (1978).
43. Sekiguchi, H. and Coutin, B. Polymerization ability and related problems in the anionic polymerization of lactams. *J. Polym. Sci. A 1.*, **11**(7), 1601 (1973).
44. Katritzky, A. R., Lomaka, A., Petrukhin, R., Jain, R., Karelson, M., Visser, A. E., and Rogers, R. D. QSPR Correlation of the Melting Point for Pyridinium Bromides, Potential Ionic Liquids. *J. Chem. Inf. Comput. Sci.*, **42**, 71 (2002).
45. Komoto, T., Iguchi, M., Kanetsuna, H., and Kawai, T. Formation of spherulites during polymerization of lactams. *Macromol. Chem.*, **135**(1), 145 (1970).
46. Kamimura, A. and Yamamoto, S. An Efficient Method to Depolymerize Polyamide Plastics: A New Use of Ionic Liquids. *Organic Letters.*, **9**(13), 2533 (2007).

47. Anderson, J. L., Ding, J., Welton, T., and Armstrong, D. W. Characterizing Ionic Liquids On the Basis of Multiple Solvation Interactions. *J. Am. Chem. Soc.*, **124**(47), 14247 (2002).
48. Sanes, J., Carri´on, F. J., Jim´enez, A. E., and Berm´udez, M. D. Influence of temperature on PA 6–steel contacts in the presence of an ionic liquid lubricant. *Wear.*, **263**, 658 (2007).
49. Jim´enez, A. E., Berm´udez, M. D., Carri´on, F. J., and Mart´ınez-Nicol´as, G. Room temperature ionic liquids as lubricant additives in steel–aluminium contacts. Influence of sliding velocity, normal load and temperature. *Wear.*, **261**, 347 (2006).

12 A Note on Nanodimensional Organosiloxanes

*B. A. Izmailov, E. N. Rodlovskaya,
and V. A. Vasnev*

CONTENTS

12.1 INTRODUCTION

The problem of microbiological stability of polymeric and textile materials in different climatic areas is actual enough for the goods, which are used in humid warm climate.

It is known [1-3], that materials and goods in air atmosphere are subjected to the influence of microorganisms such as microscopic mushrooms, actinomycete, bacteria, water plants, and so on.

The textile materials can be often enough damaged [3].

A textile material damaged by microorganisms is losing its strength; disintegrate into separate layers, crumbles.

When working on the creation of the ways to give special properties to the polymeric materials with the help of grafted micro/nanosized polyorganosiloxanes coatings [4-9], we have developed simple methods of synthesis of biocide polyorganosiloxanes layers with 1,6 di(guanidinhydrochloride)hexane groups.

12.2 EXPERIMENTAL

Polyorganosiloxane coatings with 1,6-di(guanidinhydrochloride)hexane groups have been synthesized by the molecular assembly method in two stages. At the first stage, the immobilization of oligo(chloroalkyl)ethoxisiloxane (I-IX at scheme 1) has been made on the surface of fibers by treating them with a solution in an organic solvent or with the water emulsion of oligomer (I-IX) with a specified 0,01, 0,1, 1,0, 3,0% concentration, with air-drying; after that, the modifier was fixed by 100°C heat-treating during 10 min or it was kept in the air at the room temperature during 24 hr. The characteristics of oligomers are given in the chapter [10].

As the result of the above mentioned treating, the modifier (I-IX) was covalent fixed on the surface of the material because of the condensation of ethoxygroups of the modifier with the functional groups of polymer material while building grafted micro/nanosized polyorganosiloxane coating on the surface (Scheme 1).

Scheme 1

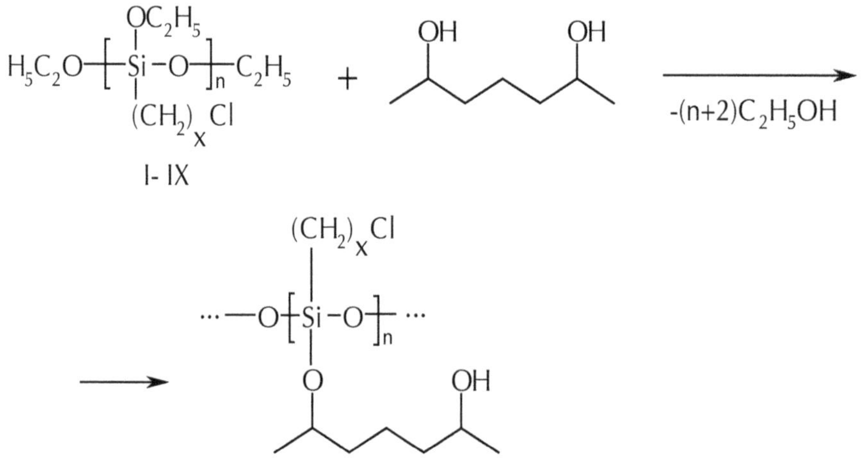

where n = 5, x = 1 (I), 3 (II), 4 (III);
n = 10, x = 1 (IV), 3 (V), 4 (VI);
n = 15, x = 1 (VII), 3 (VIII), 4 (IX).

The amount of polyorganosiloxane coating on the surface of the material have been defined after the impregnating, drying, and thermal treatment according to the increased weight of material expressed in percentages from the initial mass of the material. If after one time impregnation, drying, and thermal treatment of the textile material the increased weight did not reach the required value, then the impregnation, drying, and thermal treatment of the textile materials have been conducted some times more and it has been continued until an increased weight with the required values was reached.

At the second stage, the condensation of the grafted poly(chloroalkyl)organoxsiloxane coating with 1,6-di(guanidinhydrochloride)hexane in a alcohol solution with the presence of alkali at the room temperature was carried out.

12.3   DISCUSSION

As the result of a two staged treatment, the grafted coatings materials, which contain 1,6- di(guanidinhydrochloride)hexane groups (Scheme 2) were obtained.

Scheme 2

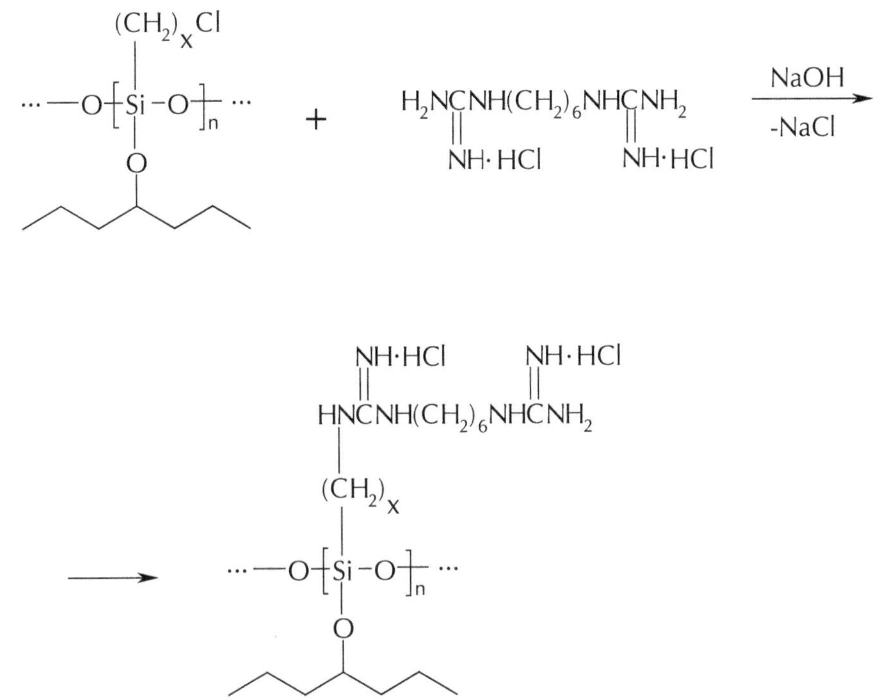

The presence of guanidine groups in the coating gives to the material a high biocide activity.

Such coatings are very effective for bacteria *E. coli*, *P. Aeruginosa*, fungi *Penicillium chrysogenum*, *Aspergillus niger*, yeast spores *Saccharomyces cerevisae*, as well as for other bacteria, fungi, and yeast spores.

The resistance of the textile samples to microbiologic destruction after their treatment with 1,6-di(guanidinhydrochloride)hexane has been examined with using of soil method (GOST-9.060-75), according to the index of microbiologic destruction resistance coefficient (Table 1).

The essence of the method is that the models in defined conditions are subjected to the influence of a natural complex of soil microflora by putting it onto the surface of the textile sample and then the coefficient of stability (%) to microbiologic destruction is defined as the relationship of treatments in case of breaking of prototype and of initial sample (without being exposed to microbiologic influence). The textile material is considered stable if the coefficient value ≥ 80% is reached.

The results indicated in the Table 1 show that a textile sample, which is modified by polyorganosiloxane coating with 1,6-di(guanidinhydrochloride)hexane groups in the amount of 0.1–1.0% mass., is almost not subjected to any microbiological destruction: The coefficient of stability makes 85–97% depending on the structure of polyorganosiloxane coating and on the amount of washings. The textile sample keeps its antimicrobial properties after 15 soap-and soda treatments conducted in accordance with the standard GOST 12.4.049-87.

TABLE 1   Influence of the composition of the modifying polyorganosiloxane coating and of washings amount on the coefficient of stability to microbiological destruction of cotton textile samples (coarse calico, art.262-063).

Coating		Coefficient of antimicrobial stability of fabric before/and after washings, %			
Oligomers	Amount, %-mass.	Before washing	After 5 washings	After 10 washings	After 15 washings
I	0,1	88,2	87,5	87,0	86,0
	1,0	95,6	95,0	94,0	94,0
II	0,1	88,5	88,0	87,0	87,0
	1,0	97,0	96,0	96,7	96,4
III	0,1	89,4	88,7	88,0	87,0
	1,0	96,8	96,0	95,8	95,5
IV	0,1	90,1	89,5	89,0	88,4
	1,0	98,3	98,0	97,2	97,0
V	0,1	92,0	91,5	91,0	90,1
	1,0	99,0	98,0	97,5	97,0
VI	0,1	89,1	88,7	88,0	87,0
	1,0	95,5	95,3	94,7	94,3
VII	0,1	90,3	88,0	87,0	85,0
	1,0	97,0	96,8	96,3	95,3
VIII	0,1	89,0	88,0	85,0	83,0
	1,0	96,8	96,5	96,0	95,5
IX	0,1	88,0	87,0	86,0	85,0
	1,0	98,2	98,0	98,0	97,3
Reference sample	0	50	–	–	–
	0	50	–	–	–

So, as a result of the conducted investigations, it was established that for protection of cotton textile against biodestruction, their fiber should have not less than 0.1% mass of organosiloxane coating with 1,6-di(guanidinhydrochloride)hexane groups.

12.4   CONCLUSION

New approaches and principles of the creating of layered micro/nanosized functional polysiloxan coatings of the given structure, composition, and texture, which are immobilized on the surface of materials allow enhancing the efficiency of the practical usage of such materials, to improve their quality and field-performance data.

KEYWORDS

- **1,6 Di(guanidinhydrochloride)hexane groups**
- **Microbiological**
- **Microorganisms**
- **Poly(chloroalkyl)organoxsiloxane**
- **Polyorganosiloxanes**

REFERENCES

1. Izmailov, B. A. and Gorchakova, V. M. Operating characteristics enhancement of non-woven materials by means of layered nano-dimensional organosiloxane coatings. *Non-woven materials*, (1), 18–21 (2007).
2. Izmailov, B. A. and Gorchakova, V. M. Defense of textile materials from biodeterioration and moisture. *Non-woven materials*, **1**(2), 10–12 (2008).
3. Izmailov, B. A. and Gorchakova, V. M. Coloration and hydrophobic trimming of non-woven materials by means of ecologically-friendly chromophoreous chlorophyll derivatives. *Non-woven materials*, **3**(4), 2–6 (2008).
4. Izmailov, B. A., Gorchakova, V. M., and Vasnev, V. A. Novel effective fibrous sorbate for non-woven materials. *Non-woven materials*, **4**(9), 38–40 (2009).
5. Izmailov, B. A. Synthesis and properties of fibrous sorbents with imparted nano-dimensional organosiloxanes the polymeric coverings containing aminomethylenphosphonic ligands. The Bulletin of the Moscow state textile university, 55–58 (2010).
6. Izmailov, B. A., Vasnev, V. A., Keshtov, M. L., Krayushkin, M. M., Shimkina, N. G., Barachevskii, V. A., and Dunaev, A. A. Photochromic silicon polymers based on 1,2-dihetarylethenes. *Polymer Science*. Series C., **51**(1), 51 (2009).
7. Andrenjuk, E. I., Bilaj, V. I, Koval, E. Z., Kozlova, I. A. Microbic corrosion and its activators. *Kiev. Scientific thought*, 288, (1980).
8. Bobkova, T. S., Zlochevskaja, I. V., Rudakova, A. K., and Cherkunova, L. N. Damage of industrial materials and products under the influence of microorganisms. *M.: MSU*, 25–31 (1971).
9. Kanevsky, I. G. Biological damage of industrial materials. *L. Science*, 232 (1984).
10. Izmailov, B. A. Designing on a surface of fibrous materials aminomethyl(organo)siloxanic coatings from functional olygomeric precursors. The Bulletin of the Moscow state textile university, 84–88 (2007).
11. Vointseva, I. I. and Gembitsky, P. A. *Poliguanidiny—disinfection means and multifunctional additives in composite materials*. M.: OOO "LKM-PRESS", 303 (2009).

13 A Note on Polythienophenes: Synthesis and Characterization

E. N. Rodlovskaya, V. A. Vasnev, and B. A. Izmailov

CONTENTS

13.1 INTRODUCTION

Polyconjugated sulfur-containing polymers, first of all polytiophene are replacing substances and analogous substances attract the attention of researchers because of the wide spectrum of their properties. Electroconductivity, luminescence, and electroluminescence condition the application of these polymers as organic conductors and semiconductors, as light emitting diodes, as photo voltage cells, sensors, photodetectors, and so on [1-5]. There are lots of articles summarized in the reviews [5-10], devoted to the chemistry of polythiophene of its co-polymers and block-co-polymers. The data about the electropolycondensation of thiophene [11, 12] and also of organoboron [13] and organometallic [14] thiophene containing polymers were published. In this review, the most important progress in the synthesis of tienothiophene containing and of dithienothiophene containing polymers during the recent 10 years is represented.

13.2 BACKGROUND

Among the thienothiophene structures [15] ones differ thieno[2,3-b]thiophene (I), thieno[3,2-b]thiophene (II), and thieno[3,4-b]thiophene (III) и thieno[3,4-c]thiophene (IV), which is known as tetraphenyl derivative of one as follows:

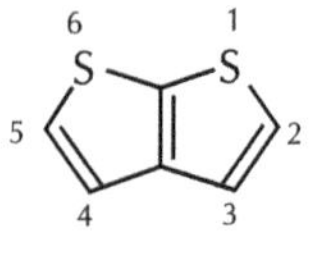

I 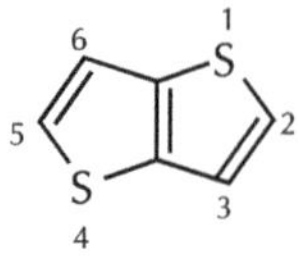   II 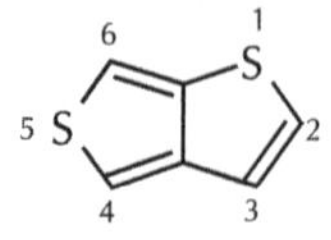   III 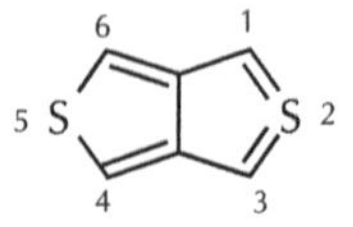   IV

In case of dithienophenes ones differ [8] dithieno [3,2-b;2',3'-d] thiophene (V), dithieno [3,4-b;3',4'-d] thiophene (VI), dithieno [2,3-b;3',2'-d]thiophene (VII), dithieno[2,3-b;2',3'-d]thiophene (VIII), dithieno [3,4-b;3',2'-d] thiophene (IX), and dithieno [3,4-b;2',3'-d] thiophene (X) as follows:

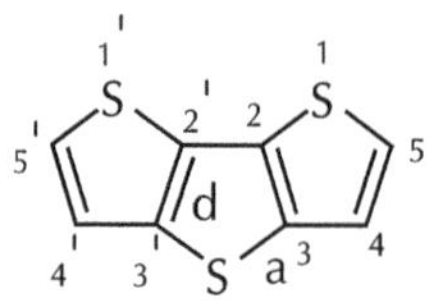

V 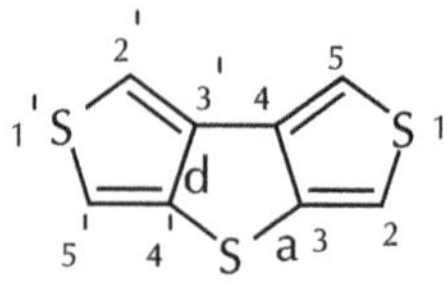   VI 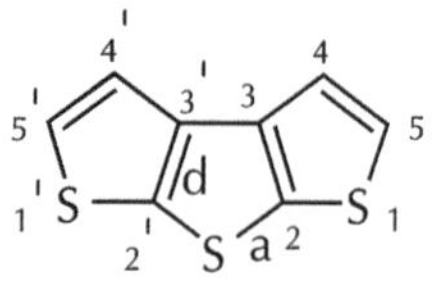   VII

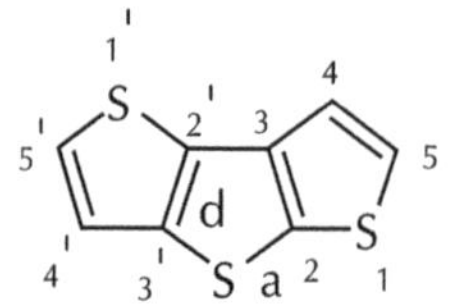

VIII 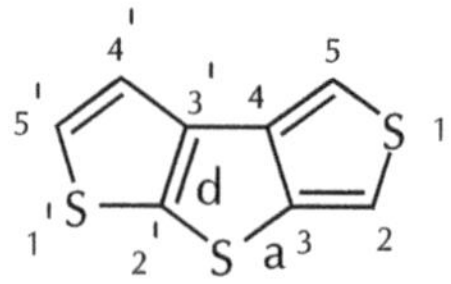   IX 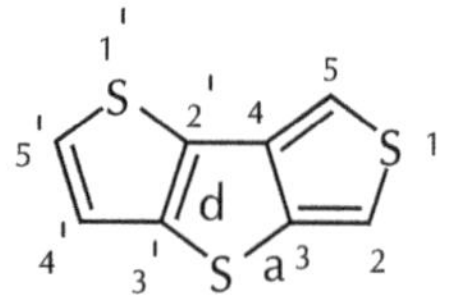   X

Polythienothiophenes are a special class of π-conjugated olithienothiophene, which are notable for thermal stability, for convertibility, and mechanical stability. To that group belong polymers or co-polymers, containing thiophene cycles, which are directly annelered to the second thiophene ring.

The quantum chemical accounts have shown [16] that the nearby degeneracy energy of aromatic form in polymers of the type XI exceeds the degeneracy energy of quinoid form XII by 1,3 kilo joule only as follows:

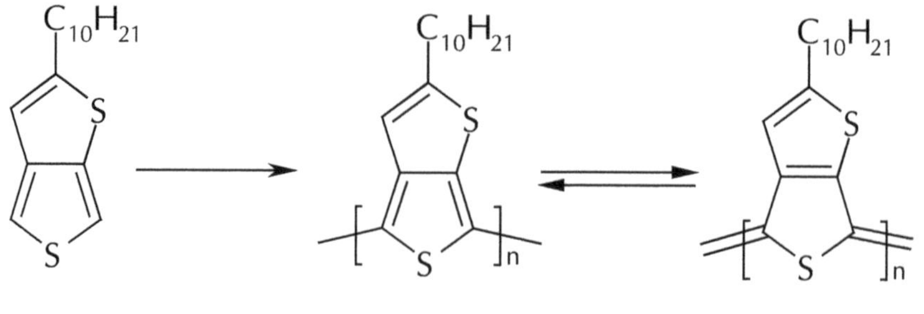

XI   XII

Therefore, the thiophenanneled polymers, apparently, can be partly in quinoid form and that is why the degree of their conjugation is growing.

The synthesis of polythienothiophenes XI has been made both by oxidizing dehydropolycondensation in chloroform with iron chloride (3^+) [17], and electrochemically in acetonitrile on a platinum electrode [18, 19]. The alkyl radical had been introduced into the polymer chain to improve the solubility of XI. Another way to enhance the solubility was the introduction of different groups, for example, a carboxyl group [20] or of perfluoroalkyl groups [21] into the initial thienothiophene monomer.

With the electropolymerization, the soluble polythienothiophen XIII having satisfactory electrophysical parameters has been obtained [22]. The band gap of resulting polymers varied from 0.78 to 1.0 eV as follows:

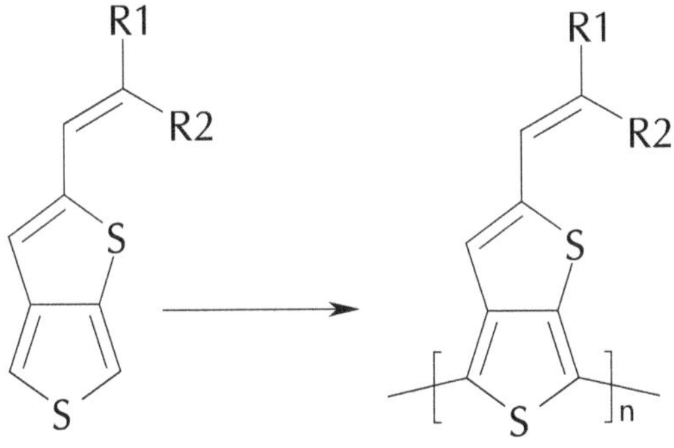

XIII

$R_1 = C_5H_{11}$, $R_2 = H$; $R_1 = Ph$, $R_2 = H$; $R_1 = Ph$, $R_2 = CN$.

For the purpose of improving electro-physical characteristics (small band gap ~1.1 eV), polydithienothiophenes have been synthesized XIV-XVI [23, 24] as follows:

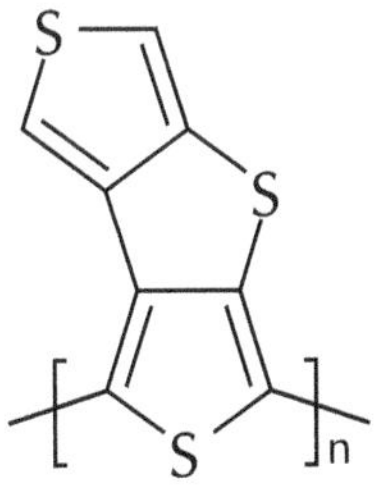

XIV 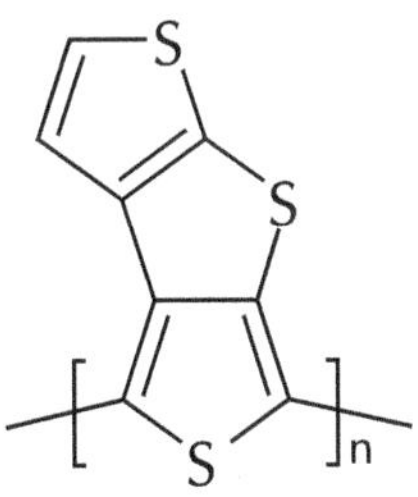   XV 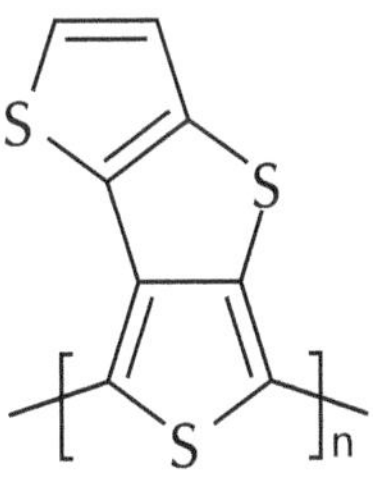   XVI

It should be noted that with the increase of the molecular mass, the solubility and mechanical properties of polymers become worse.

A similar reaction was obtained, polythieno[3,2-b]thiophene XVII [25-28] and co-polymer XVIII on its ground [29-32] as follows:

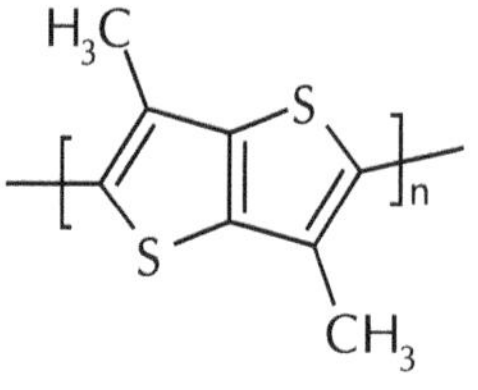

XVII

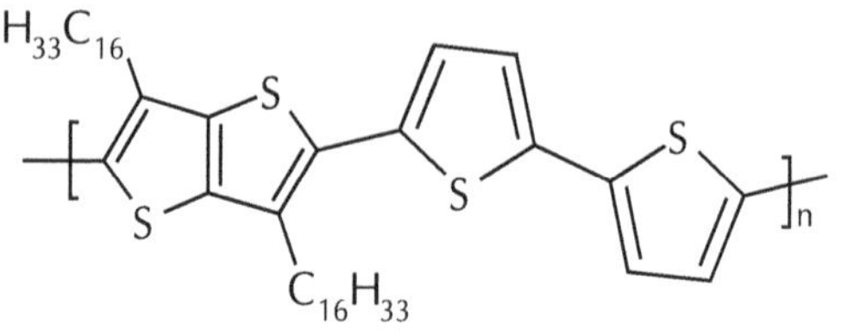

XVIII

Through the Diels–Alder reaction of cycle-adding fluorine containing copolymers of thienothiophene XIX has been obtained where the effect of self-organization is very high. The authors attribute this with the presence of fluorine atoms in polymers [33] as follows:

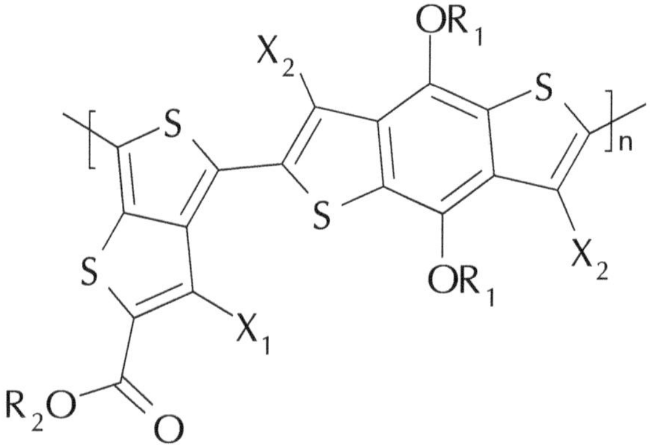

XIX

X_1 = H, X_2 = H; X_1 = F, X_2 = H; X_1 = H, X_2 =F; X_1 = F, X_2 = F.

The soluble, conjugated, and alternated thienothiophene containing co-polymers XX have been obtained as result of the reaction Stille, catalyzed with palladium(2^+) [34-36] as follows:

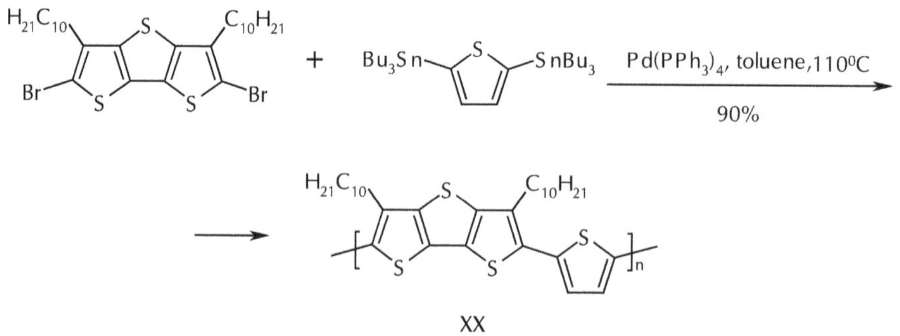

XX

The new type of donor-acceptor co-polymers XXI has been synthesized by the reaction of olefination Horner Wadsworth Emmons, where the thienothiophene fragments are acting as donor and benzoxasole ones as acceptor [37] as follows:

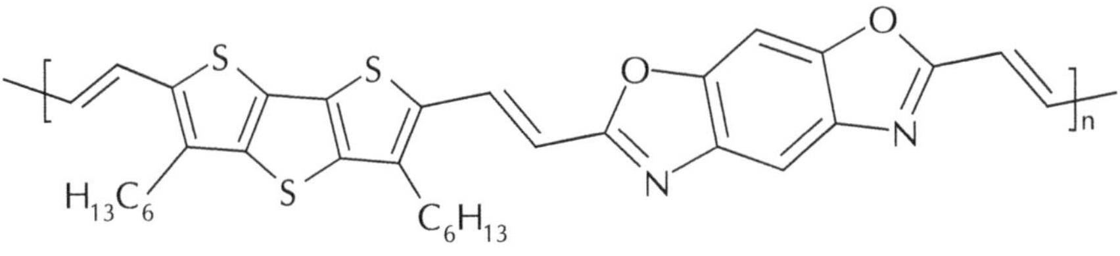

XXI

Processes of a deep sulfurization of polyethylene [38, 39], polystyrene [40], poly-vinylpyridine [41], and polyvinylchloride [42] have been investigated. It was found that these polymers can be completely sulfurized by elemental sulfur forming thio-phene containing polymers XXII and XXIII as follows:

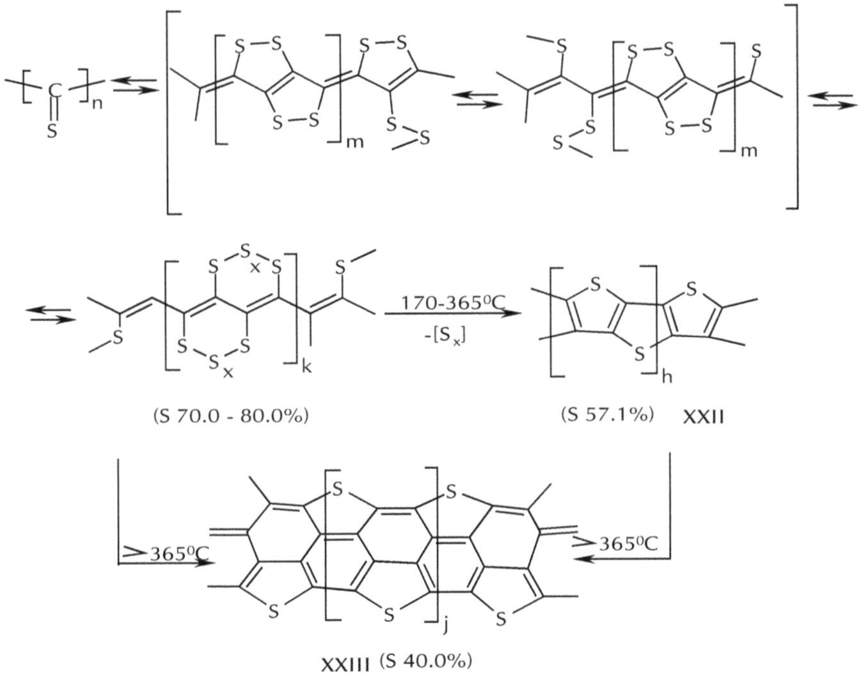

(S 70.0 - 80.0%)   170-365°C   (S 57.1%)   XXII
                                      -[S$_x$]

XXIII (S 40.0%)

Believe that at the first stage of the process at relative low temperatures (~200°C) carbon polysulphide is generated, and then from one are formed polyen-polysulphide fragments with a high sulfur content (~70%); then these fragments subjected to desul-furization forming more stable polythienothiophene structures XXII and of parquet polynaphtothienothiophene forms XXIII. Prepared polymers represent black bright powders, having electroconductivity and paramagnetism.

Recently, the synthesis of polythiophene structures through polymer-analogous reactions becomes more and more significant. So, one of the authors has made the synthesis of polythienothiophenes with ketone XXIV and amide XXV bridging groups [43, 44], according to the scheme as follows:

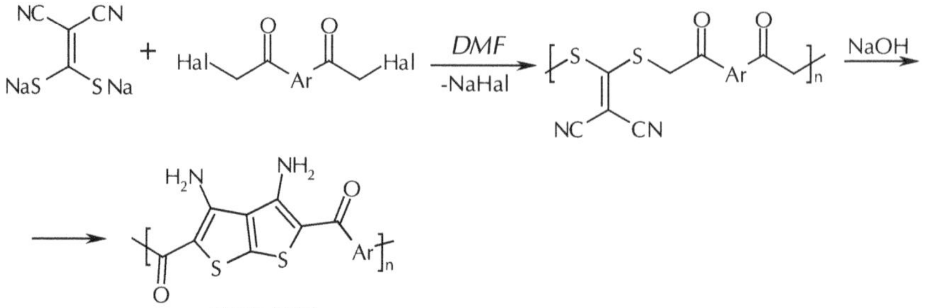

XXIV, XXV

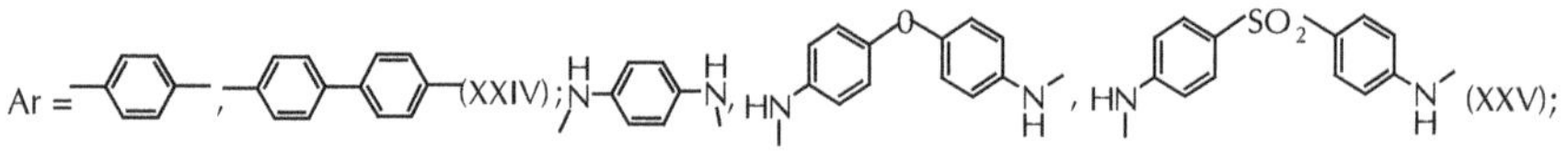

Hal = Br, Cl.

The thiophene fragments in such type of polymers are forming as a result of izomerization polymer-analogous transformation directly during the synthesis.

13.3   CONCLUSION

The analysis of data shows that an intensive development of polythienothiophene chemistry is caused by the complex of new photo and electrochemical properties of sulfur containing polymers.

The most widely used methods of the synthesis of thiophen-containing polymers, based on the using of monomers containing a ready thiophene ring.

The work is fulfilled with financial support of Russian fund of fundamental investigations (code of the design is 11-03-00426).

KEYWORDS

- **Co-polymers**
- **Diels–Alder reaction**
- **Electropolymerization**
- **Monomers**
- **Polythienothiophene**
- **Synthesis of polythiophene**
- **Thiophene ring**

REFERENCES

1. *Conjugated Polymers: The Novel Science and Technology of Highly Conducting and Nonlinear Optically Active Materials.* J. L. Bredas and R. Silbey (Eds.). Kluwer Academic, Dordrech (1991).
2. *Handbook of Oligo and Polythienothiophenes.* D. Fichou (Ed.). Wiley-VCH, Weinheim (1999).

3. *Applied Polymer Science.* C. D. Craver and E. Charles (Eds.). Pergamon Press, Oxford (2000).
4. *Handbook of Nanostructured Materials and Nanotechnology.* H. S. Nalwa (Ed.). Academic Press, New York (2000).
5. Perepichka, I. F., Perepichka, D. F., Meng, H., and Wudl, F. Light-Emitting Polythiophenes. *Adv. Matter.,* **17**, 22812305 (2005).
6. Janosik, T. and Bergman, J., Chapter 5.1: Five-membered ring systems: Thiophenes and Se/Te analogs. *Progr. Heterocycl. Chem.,* **20**, 94121 (2009).
7. Rodlovskaya, E. N., Frolova, N. G., Savin, E. D., and Nedel'kin, V. I. Achievements in the Synthesis of Thiophene-containing Polymers. *Polymer Science, Ser. C,* **48**(1), 5884 (2006).
8. Ozturk, T., Ertas, E., and Mert, O. *Dithienothiophenes. Tetrahedron,* **61**, 1105511077 (2005).
9. Yagci, Y. and Toppare, L. Electroactive macromonomers based on pyrrole and thiophene: A versatile route to conducting blocks and graft polymers. *Polym. Int.,* **52**(10), 15731578 (2003).
10. Khan, M. A. and Armes, S. P. *Conducting polymer-coated latex particles. Adv. Mater.,* **12**(9), *671674 (2000).*
11. Gurunathan, K., Murugan, A. V., Marimuthu, R., Mulik, U. P., and Amalnerkar, D. P. Electrochemically synthesized conducting polymeric materials for applications towards technology in electronics, optoelectronics and energy storage devices. *Mater. Chem. Phys.,* **61**, *173191 (1999).*
12. Wolf, M. O. Transition-metal-polythiophene hybrid materials. *Adv. Mater.,* **13**, *545553 (2001).*
13. Matsumi, N. and Chujo, Y. A new class of π-conjugated organoboron polymers. *Spec. Publ. Roy. Soc. Chem.,* **253**, 5158 (2000).
14. Kern, J. M., Sauvage, J. P., Bidan, G., and Divisia-Bloorn, B., Transition-metal-tempiated synthesis of rotaxanes and catenanes: From small molecules to poly-mers. *J. Polym. Sci.* Part A: *Polym. Chem.,***41**(22), 34703477 (2003).
15. Litvinov, V. P., and Gol'dfarb, Ya. L. The Chemistry of Thienothiophenes and Related Systems. *Adv. Heterocycl. Chem.,* **19**, 123214 (1976).
16. Hong, S. Y. and Marynick, I. D. Understanding the Conformation Stability and Electronic Structures of Modified Polymers Based on Polythiophene. *Macromolecules,* **25**(18), 46524657 (1992).
17. Pomerantz, M., Gu, X., and Zang, S. X. Poly(2-decylthieno[3, 4-b]thiophene-4,6- diyl). *A New Low Band Gap Conducting Polymer. Macromolecules,* **34**(6), 18171824 (2001).
18. Lee, K. and Sotzing, G. A. Poly(thieno[3,4-b]thiophene). A New Stable Low Band Gap Conducting Polymer. *Macromolecules.,* **34**(17), 57465747 (2001).
19. Lee, B., Seshadri, V., Palko, H., and Sotzing, G. A. Ring-Sulfonated Poly(thienothiophene). *Adv. Mater.,* **17**, 17921795 (2005).
20. Saji, V. S., Zong, K., and Pyo, M. NIR-absorbing poly (thieno[3,4-b]thiophene-2-carboxylic acid) as a polymer dye for dye-sensitized solar cells. *J. Photochem. Photobiol. A.,* **212**, 8187 (2010).
21. Shiying, Z. and Shafinq, N. F. Patent US 2010/0270055 A1, Electrically Conductive Films Formed from Dispersions Comprising Conductive Polymers and Polyurethanes., (Oct 28, 2010).
22. Park, J. H., Seo, Y. G., Yoon, D. H. Lee, Y. S., Lee, S. H., Pyo, M., and Zong, K. A concise synthesis and electrochemical behavior of functionalized poly (thieno[3,4-b]thiophenes): New conjugated polymers with low bandgap. *European Polymer J.,* **46**, 17901795 (2010).
23. Ehrenfreund, E., Cravino, A., Neugebauer, H., Sariciftci, N. S., Luzzati, S., and Catellani, M. Even parity states in small band gap p-conjugated polymers: Polydithienothiophenes. *Chem. Phys. Lett.,* **394**, 132136 (2004).
24. Neugebauer, H. Infrared signatures of positive and negative charge carriers in conjugated polymers with low band gaps. *J. Electroanal. Chem.,* **563**, 153159 (2004).
25. Nakayama, J., Dong, H., Sawada, K., Ishii, A., and Kumakura, S. Synthesis and characterization of dimers, trimers, and tetramers of 3, 6-Dimethylthieno[3,2-b]thiophene and 3,6-Dimethylselenolo[3,2-b]selenophene. *Tetrahedron,* **52**(2), 471488 (1996).
26. Diez, A. S., Saidman, S., and Garay, R. O. Synthesis of thienothiophene conjugated polymer. *Molecules,* 5, 555556 (2005).

27. Zhu, Z., and Waller, D. Patent WO 2010/016986 A1, *Novel Photoactive Co-Polymers*, (Feb 11, 2010).
28. Daley, S. P. Patent WO 2010/005622 A2, Telescoping Devices, (Jan 14, 2010).
29. Nakayama, K., Uno, M., Nishikawa, T., Nakazawa, Y., and Takeya, J. Air-stable and high-mobility organic thin-film transistors of poly(2,5-bis(2-thienyl)-3,6-dihexadecyltheino[3,2-b] thiophene) on low-surface-energy self-assembled monolayers. *Organic Electron.*, **11**, 16201623 (2010).
30. Noh, Y. Y., Kim, D. Y., Misaki, M., and Yase, K. Highly polarized light emission from uniaxially aligned thin films of biphenyl/thienothiophene co-oligomer. *Thin Solid Films.*, **516**, 7505–7510 (2008).
31. Savenije, T. J., Grzegorczyk, W. J., Heeney, M., Tierney, S., McCulloch, I., and Siebbeles, L. D. A. Photoinduced Charge Carrier Generation in Blends of Poly(Thienothiophene) Derivatives and [6,6]-Phenyl-C61-butyric Acid Methyl Ester: Phase Segregation versus Intercalation. *J. Phys. Chem. C*, **114**(35), 15116–15120 (2010).
32. Wallace, P. and Goddard, S. Patent WO 2010/018381 A1, Opto-Electrical Devices and Method of Manufacturing the Same. (Feb 18, 2010).
33. Son, H. J., Shin,W. W., Xu, T., Liang, Y. Wu, Y. Li, G., and Yu. L. Synthesis of Fluorinated Polythienothiophene-co-benzodithiophenes and Effect of Fluorination on the Photovoltaic Properties. *J. Am. Chem. Soc.*, **133**(6), 1885–1894 (2011).
34. Turbiez, M., Hergue, N., Leriche, P., and P. Frere. Rigid oligomers based on the combination of 3,6-dimethoxythieno[3,2-b]thiophene and 3,4-ethylenedioxythiophene. *Tetrahedron Lett.*, **50**, 7148–7151 (2009).
35. Paeka, S., Leea, J., Lima, H. S., Lima, J., Lee, J. Y., and Lee, C. The influence of electron-deficient comonomer on chain alignment and OTFT characteristics of polythiophenes. *Synth. Met.*, **160**, 2273–2280 (2010).
36. Zhang, S., He, C., Liu, Y., Zhan, X., and Chen, J. Synthesis of a soluble conjugated copolymer based on dialkyl-substituted dithienothiophene and its application in photovoltaic cells. *Polymer.*, **50**, 3595–3599 (2009).
37. Patil, A. V., Park, H., Lee, E. W., and Lee, S. H. Synthesis and characterization of dithienothiophene vinylene based co-polymer for bulk heterojunction photovoltaic cells. *Synth. Met.*, **160**, 2128–2134 (2010).
38. Trofimov, B. A., Skotheim, T. A., Mal'kina, A. G., Sokolyanskaya, L. V., Myachina, G. F., Korzhova, S. A., Stoyanov, E. S., and Kovalev, I. P. Sulfurization of polymers. 2. Polythienothiophene and related structures from polyethylene and elemental sulfur. *Russ. Chem. Bull.*, **49**(5), 863869 (2000).
39. Trofimov, B. A., Skotheim, T. A., Mal'kina, A. G., Sokolyanskaya, L. V., Myachina, G. F., Korzhova, S. A., Vakul'skaya, T. I., Kovalev, I. P., Mikhailik, Yu. V., and Boguslavskii, L. I. Sulfurization of polymers. 3. Paramagnetic and redox properties of sulfurized polyethylene. *Russ. Chem. Bull.*, **49**(5), 870–873 (2000).
40. Trofimov, B. A., Skotheim, T. A., Mal'kina, A .G., Sokolyanskaya, L. V., Myachina, G. F., Korzhova, S. A., Vakul'skaya, T. I., Klyba, L. V., Stoyanov, E. S., Kovalev, I. P., and Mikhailik, Yu. V. Sulfurization of polymers. 4. Poly(4,5,6,7-tetrathiono-4,5,6,7-tetrahydrobenzothiophene-2,3-diyl) and related structures based on polystyrene and elemental sulfur. *Russ. Chem. Bull.*, **50**(2), 253–260 (2001).
41. Trofimov, B. A., Mal'kina, A. G., Sokolyanskaya, L. V., Nosyreva, V. V., Myachina, G. F., Korzhova, S. A., Rodionova, I. V., Vakul'skaya, T. I., Klyba, L. V., Stoyanov, E. S., Skotheim, T. A., and Mikhailik, Yu. V. Sulfurization of polymers. 5. Poly(6-methyl-5-sulfanylthieno[2,3-b] pyridine-4-thione), poly(thieno[2,3-b]azepine-4,5(6H)-dithione), and related structures from poly(2-methyl-5-vinylpyridine) and elemental sulfur. *Russ. Chem. Bull.*, **51**(2), 282–289 (2002).
42. Trofimov, B. A., Vasil'tsov, A. M., Petrova, O. V., Mikhaleva, A. I., Myachina, G. F., Korzhova, S. A., Skotheim, T. A., Mikhailik, Yu. V., and Vakul'skaya, T. I. Sulfurization of polymers. 6. Poly(vinylene polysulfide), poly(thienothiophene), and related structures from polyacetylene and elemental sulfur. *Russ. Chem. Bull.*, **51**(9), 1709–1714 (2002).

43. Rodlovskaya, E. N., Frolova, N. G., Savin, E. D., and Nedel'kin, V. I. Synthesis of Polythienothiophenes by Polycondensation of Sodium 1, 1-Dicyanovinylenedithiolate with Bis(chloroacetanilide)s. *Polymer Science, Ser. A*, **46**(10), 1017–1021 (2004).
44. Rodlovskaya, E. N., Frolova, N. G., Savin, E. D., and Nedel'kin, V. I. Synthesis of Polythienothiophenes by Polycondensation of Sodium 1,1-Dicyanovinylenedithiolate with bis(Bromoacetyl) arylenes. *Polymer Science, Ser. A*, **46**(6), 593–598 (2004).

14 Synthesis and Behavior of Phthalide-containing Polymers

S. N. Salazkin and V. V. Shaposhnikova

CONTENTS

14.1 INTRODUCTION

Polymers containing cardo groups and related groups concern to cardo polymers of many classes [1-4] and as a rule are heterochain polymers. However, polymers in which phthalide groups and related groups are connected with one another by hydrocarbon structures without heteroatoms have been prepared as follows:

X	Y	Polymer
$-O-$	$>C{=}O$	Phthalides
$>N{-}R_1$	$>C{=}O$	Phthalimidines
$-O-$	$>S\!\!\stackrel{O}{\underset{O}{\diagup}}$	Sulfophthalides

Synthesis of such polymers was carried out by both polymerization [5] and poly-condensation [6]. Polymer synthesis by polycondensation was provided by using of new approach to carrying out of polycondensation. In last year interest to this poly-mers, especially to poly(diphenylene phthalide), has been increased due to revealing of novel valuable properties for these polymers (effect of electron switching, change of optical, and magnetic properties in result of external influences and others) [7].

14.2   RESULTS

Poly(methylidene phthalide) is prepared by polymerization of methylidene phthalide as follows:

$$ n \underset{II}{\underset{}{\left[\text{structure} \right]}} \longrightarrow \underset{III}{\underset{}{\left(\text{structure} \right)_n}} \tag{2} $$

Evidently, this polymer was prepared for the first time by S. Gabriel [8] as by prod-uct during synthesis of methylidene phthalide. As a desired product, this polymer was prepared greatly later [9] by radical polymerization of methylidene phthalide. Howev-er, pioneering systematic investigation of synthesis and properties of poly(methylidene phthalide) was carried out only in 6070 years of 20th century [10-13].

Poly(methylidene phthalide) has the highest concentration of cardo groups in its polymer chains in which cardo phthalide groups are separated from one another only by one methylene group. This peculiarity of the chemical structure imparts high rigidi-ty to poly(methylidene phthalide) macromolecules and thus makes the polymer highly heat-resistant. The promise of methylidene phthalide (co)polymers and the interest in their practical application are not only due to their heat-resistance and optical transpar-ency properties, which are of importance for traditional applications. The latest studies devoted to the properties of different phthalide-containing polymers [7] lead us to as-sume that methylidene phthalide (co)polymers may possess a set of properties typical of smart polymers that is the ability to switch over electric current and/or change opti-cal properties during external actions.

The efficient synthesis of high-purity methylidene phthalide is of key significance for improving the synthesis of methylidene phthalide-based polymers and their practi-cal use.

At present, there are two ways to synthesize methylidene phthalide: namely, the decarboxylation of phthalylideneacetic acid and the dehydration of o-acetylbenzoic acid.

(3)

The first methods of realization of these routes of methylidene phthalide synthesis (decarboxylation [8, 14] and dehydration [15, 16]) have not provided necessary quality and high yield of methylidene phthalide. Further perfection of these methods [17] (thermal decarboxylation or dehydration at simultaneous vacuum sublimation of formed monomer) leads to sharp increasing of yield but does not solved the problem of monomer quality.

Optimization [5] of dehydration of o-acetylbenzoic acid in the system N,N-dimethylformamid (DMF)-$SOCl_2$ [17] provides relatively successful synthesis of methylidene phthalide (high yield and grade).

The high quality of the monomer ensures good and reproducible results for its polymerizaton, which yields polymer with high molecular mass and good properties [5].

For this purpose, it is necessary to take into account our data on the influence of water on the polymerization of methylidene phthalide, which readily binds up to 0.5 mol of water [19]. The monomer must be free of chlorine-containing impurities, while the content of allowable water is insignificant ($<<0.5\%$). In this case, thermal- and peroxide-induced (co)polymerization of methylidene phthalide may be successfully performed in both bulk and solution. The polymerization of methylidene phthalide in DMF (4 g of the monomer in 10 ml of the solvent, initiation with benzoyl peroxide (0.1%), 60°C) yielded poly(methylidene phthalide) with a reduced viscosity (η_{red}) equal 0.86 dl/g (DMF), and a weight-average molecular mass (Mw)1 ′ 10^5 (light scattering). Poly(methylidene phthalide) with $\eta_{red} > 1.1$ dl/g (DMF) can be easily obtained via solution polymerization by varying reaction conditions. Bulk polymerization yields poly(methylidene phthalide) with $\eta_{red} = 0.8$–2.5 dl/g (DMF). In some cases, an insoluble polymer is formed.

Now, it is reasonable to compare the main properties (primarily, the thermal stability, heat resistance, and solubility) of poly(methylidene phthalide) and poly(diphenylene phthalide) (considered in detail below) in which phthalide groups are linked via the simplest fragments, that is methylene groups or diphenyl fragments. Furthermore, this comparison is useful because the properties of poly(diphenylene phthalide) that make it possible to attribute the latter to smart polymers have been investigated in detail [20].

Poly(methylidene phthalide) is readily soluble in strongly polar protic and aprotic solvents, such as trifluoroacetic acid, 85% formic acid, dimethylsulfoxide (DMSO), 1,1,3,3-tetrafluoro-1,3-dichloroacetone monohydrate, hexafluoroisopropanol, thrifluoroethanol, pentafluoromonochloroacetone and hexafluoroacetone sesquihydrates, concentrated sulfuric acid, sulfolane, N-methylpyrrolidone, DMF, and tricresol (the solvents are arranged approximately in the order of decreasing strength). It seems that only trifluoroacetic acid can be attributed to good solvents (on the basis of intrinsic viscosity $[\eta]$ and the second virial coefficient, see Figures 1(a)–(c). Poly(methylidene phthalide) is insoluble in benzene in which polystyrene is readily soluble (although polystyrene is insoluble in trifluoroacetic acid, 85% formic acid, and DMF). This difference from polystyrene obviously demonstrates the effect of phthalide groups on solubility. Poly(methylidene phthalide) is likewise insoluble in chloroform and symm-tetrachloroethane in both of which poly(diphenylene phthalide) is readily soluble.

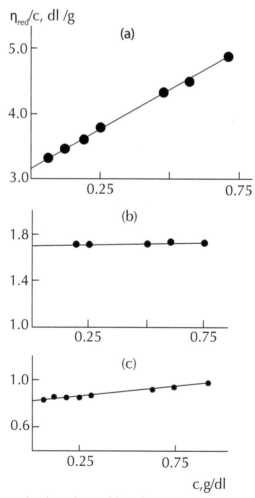

FIGURE 1 Concentration dependence of the reduced viscosity of poly(methylidene phthalide) solutions in (a) trifluoroacetic acid, (b) DMSO, and (c) DMF. The polymer was obtained in DMF at 60°C in the presence of benzoyl peroxide (0.1 wt %) for 12 hr.

The thermal stability of poly(methylidene phthalide) is governed by its depolymerization ability, which is inherent in polymerization-synthesized polymers. Depolymerization in vacuum is detected at an initial value of 275°C, while the intense process is observed at 325°C. In air, this process proceeds lower at temperatures 25°C (The onset temperature of depolymerization is 250°C, while the intense process develops at 300°C). In this respect, poly(diphenylene phthalide) ranks above poly(methylidene phthalide) because it is not subjected to depolymerization and its onset temperature of mass loss is about 440–460°C.

The glass transition temperature of poly(methylidene phthalide) must be approximately 380–390°C (according to estimates, 400–410°C) [12]. This result implies that in terms of this characteristic, poly(methylidene phthalide) must be comparable to poly(diphenylene phthalide), whose T_g is 420°C. However, this heat-resistance potential of poly(methylidene phthalide) cannot be realized, because its depolymerization begins at a substantially lower temperature (below 300°C) than its supposed glass transition temperature. The monomer resulting from depolymerization (or contained in poly(methylidene phthalide) as a result of incomplete bulk polymerization) plasticizes the polymer, and the T_g of poly(methylidene phthalide) observed in the tests never reaches 300°C ($Tg = 215$–240°C, depending on the conditions of sample preparation, that is bulk polymerization under different regimes or molding of polymer isolated after polymerization and washed to remove the monomer). In this situation, it may be more reasonable to use methylidene phthalide for copolymerization with well known common monomers (styrene, methyl methacrylate, methyl acrylate, acrylonitrile, etc.). For example, methylidene phthalide copolymers with styrene, methyl methacrylate, acrylonitrile, and methyl acrylate with $T_g = 220$–260°C have been obtained. In the case of the three former comonomers, a content of methylidene phthalide units equal to 40 mol % is sufficient for the synthesis of copolymers with $T_g > 200$°C. The following fact is of principle significance. Copolymers with very low contents of styrene (10–20%), methyl methacrylate (10 mol %), and methyl acrylate (10 mol %) have T_g ³ 260°C (for styrene copolymers, 270°C). This circumstance suggests that the incorporation of the above comonomers into the copolymers somewhat increases, rather than decreases, the real glass transition temperature. This finding is likely related to the fact that the synthesis of copolymers leads to an increase in stability against depolymerization. This assumption is supported by the data on the thermal stability of some copolymers. The transparency of the block resulting from bulk copolymerization is attained only at specified comonomer ratios. Methylidene phthalide copolymers are transparent when the contents of styrene, methyl methacrylate, and methyl acrylate contents are no higher than 30 mol % and the content of acrylonitrile content is no higher than 50 mol %.

Poly(methylidene phthalide) has a low shock resistance (the reduced impact strength of blocks is 1–2 kJ/m²). Such a high brittleness of poly(methylidene phthalide) is obviously due to a very high concentration of phthalide groups in its macromolecules. In this respect, poly(diphenylene phthalide) has a better set of characteristicsIts pressed specimens have satisfactory resistance to shock loads (the reduced impact strength is 7–8 kJ/m²), and its films are characterized not only by high strength but also, in contrast to brittle poly(methylidene phthalide) films, by elasticity (They are highly resistant to repeated bends). The synthesis of copolymers with comonomer contents below 30 mol % does not eliminate the brittleness of bulk specimens of such copolymers. (Their impact strength is 1–3 kJ/m².) At present, only methylidene phthalide copolymers with acrylonitrile containing 50 mol % of the latter monomer have satisfactory impact strength (7–8 kJ/m²); after modification of these copolymers with small amounts of cross-linking agents, their impact strength increased to 11 kJ/m².

It is proposed that poly(methylidene phthalide) by analogy with polydiphenylenephthalide [6] will possessed by properties, which permit to relate the polydiphenylenephthalide to "smart" polymers [7, 20, 21].

It is necessary to note on poly(methylidene phthalimidine) (that is polymer containing phthalimidine group instead of related phthalide group) among polymers related to poly(methylidene phthalide). Synthesis of such polymer [22] based on effective preparation of high pure monomer [23] may be presented by the following scheme:

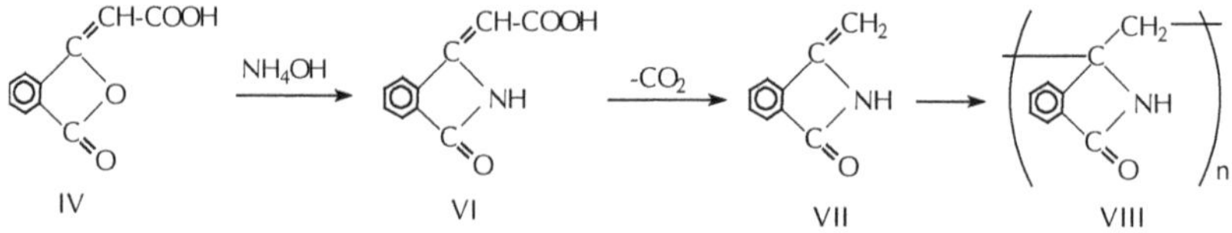

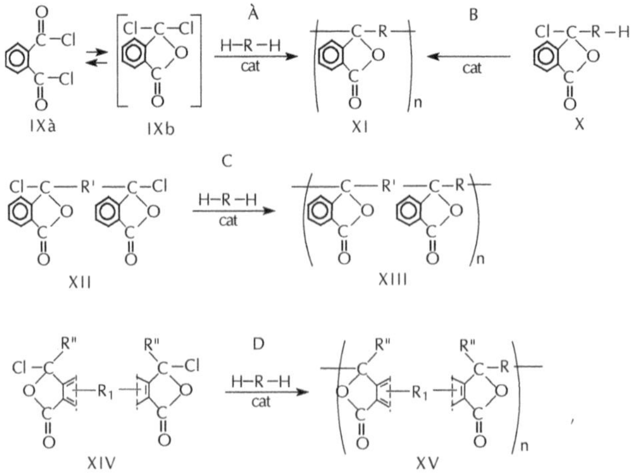

Polycondensation by mechanism of electrophilic substitution reaction at using as monomer pseudoacid chlorides of α-ketocarbonic acid is in principle new approach to synthesis of phthalide-containing aromatic polymers. In this case, the growth of a macromolecular chain proceeds through the formation of a carboncarbon bond directly between an aromatic nucleus and a carbon atom of the phthalide cycle; this bond results from a reaction with the aromatic nucleus of the functional group (chlorine) occurring immediately at the carbon atom of the phthalide cycle. The first communications on the synthesis and properties of polyarylenephthalides [24-29] revealed that this new line of research holds much promiseThe possibility of preparing high molec-ular-mass polyarylenephthalides lacking a cross-linked structure was established, and the valuable properties of these polymers (high thermal stability and good heat and chemical resistance) were demonstrated. Synthesis of polyarylene phthalides is pos-sible by several variants presented in the following scheme [24, 25, 27-29]:

Here H-R-H и H-R'-Hare aromatic polynuclear hydrocarbons as follows:

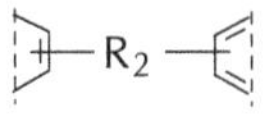

R"are aromatic radicals, for example, phenyl.

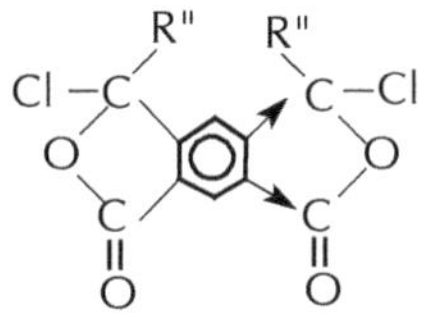

Further studies have mostly focused on variants (B) and (C), where R and/or R' = R_3 = H, alkyl

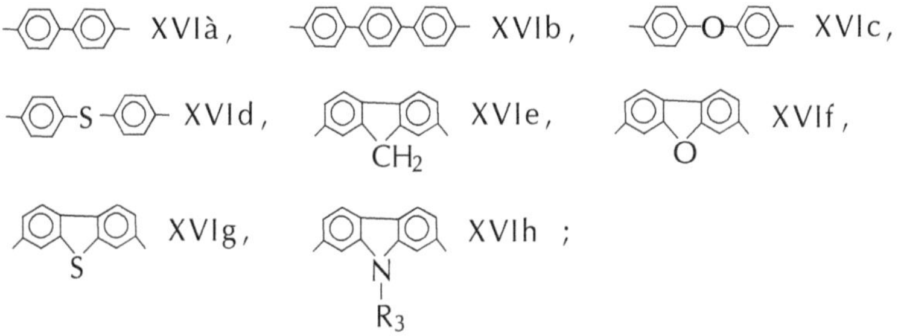

In the case of R and R' = XVIa, XVIb и XVIc polyarylenephthalides concerning to interesting us in this chapter polymers type, namely, polymers in which phthalide groups are connected with one another by carbohydrate structures without heteroat-oms have been prepared.

Variants (B) and (C) exhibit the most promise for polymer synthesis, since the ob-taining of corresponding monomers and intermediates involves well studied reactions with the use of accessible reagents and rather simple experimental techniques [30, 31]. The synthesis of these intermediates and monomers may be presented by the general schemes as in the following:

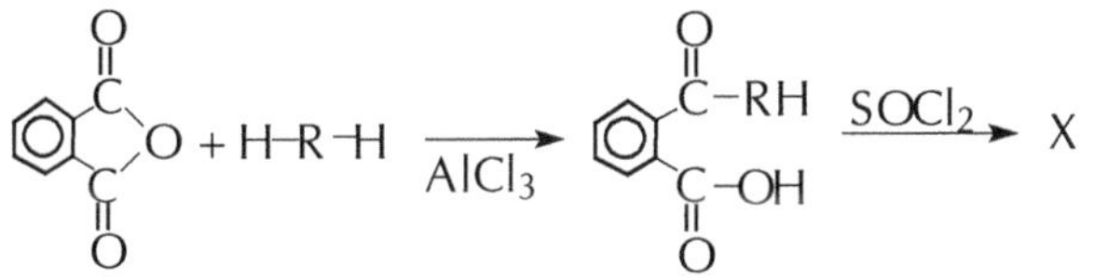

$$(6)$$

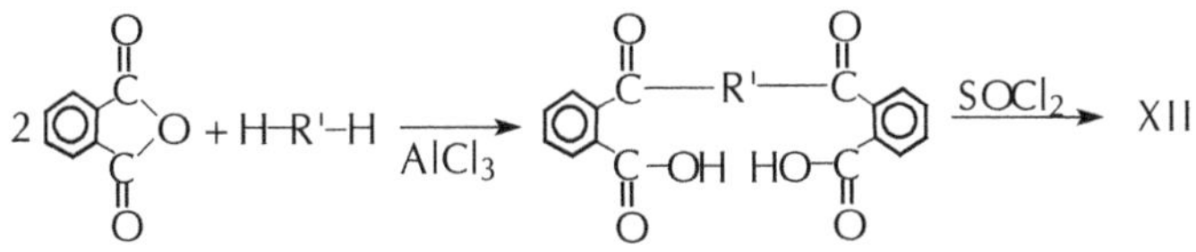

$$(7)$$

H–R–H and H–R'–H are aromatic polynuclear hydrocarbons

Very detailed information is available for the synthesis of polyarylenephthalides by variant B in Scheme (8), that is homopolycondensation (self-condensation) of pseudoacid chloride of mono-α-ketocarboxylic acids, especially the homocondensation of 3-chloro-3-(diphenyl-4'yl)phthalide, which affords one of the most valuable polymers polydiphenylenephthalide (or poly(3,3-phthalylidene-4›,4»-diphenylene).

$$\text{XVII} \xrightarrow{\text{cat}} \text{XVIII} \tag{8}$$

In [29, 32-35], considerable attention was given to the impact of conditions of polycondensation *via* Scheme (11) on molecular mass and selectivity, that is on the preclusion of formation of side structures, including branched and cross-linked ones. The effects of the solvent nature (chlorinated hydrocarbons and nitro compounds), the catalyst type (ZnCl$_2$, AlCl$_3$, AlBr$_3$, InCl$_3$, TiCl$_4$, SnCl$_4$, SbCl$_3$, SbCl$_5$, SbF$_5$, and FeCl$_3$), the monomer concentration (0.5–3.0 mol/l), the catalyst amount (0.5200 mol % based on the monomer), temperature (40180°C), and the time of reaction (0.2550 hr) were investigated in depth. Some model reactions were examined, a number of model compounds were synthesized, and methods for estimating the content of the most probable side products (including inactive end groups responsible for termination of chain growth) in polydiphenylenephthalide were elaborated [36-40]. The regularities of polycondensation and the effect of the parameters under consideration show a complicated pattern; these studies were detailed in [32, 33]. The AlCl$_3$ was found to be the most versatile among the tested catalysts. This catalyst exhibits catalytic activity when both chlorinated hydrocarbons (methylene chloride, sym-dichloroethane, and sym-tetrachloroethane) and nitro compounds (nitromethane, nitrobenzene, etc.) are used as solvents. However, in order to prepare high molecular-mass polymers, more than 100 mol % AlCl$_3$ is required (the optimum quantity is nearly 140 mol %). Amongst the studied solvents, nitrobenzene is the most universal. In the case of this solvent, all the foregoing catalysts are catalytically active; however, for the majority of these catalysts, their optimum amounts are high (≥100 mol % catalyst per acid chloride group) at an optimum temperature of 80–110°C. Over this temperature range, several catalysts (InCl$_3$, SnCl$_4$, SbCl$_5$, and SbF$_5$) exhibit catalytic activity that is sufficient for formation of highmolecular-mass polymers at rather small amounts of the catalysts (0.510 mol %). The polycondensation catalyzed by InCl$_3$ and, especially, SbCl$_5$ is the best-studied case.

A deeper insight into the idea of polycondensation may be provided by studying the effect of polycondensation conditions, not only on molecular mass (or η$_{red}$) [29, 33], but also on diverse chemical reactions occurring in the course of polycondensation [32]. This was achieved due to the special investigation of side reactions and the development of methods applicable to the quantitative estimation of the content of side structures and end groups of various types [36-40]. The data collected may be presented by the following scheme, which offers a complete enough notion of the complexity of polydiphenylenephthalide synthesis as follows:

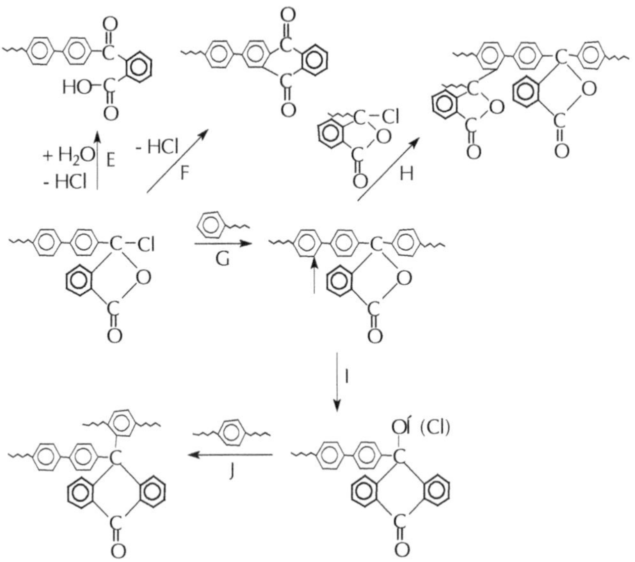

$$(9)$$

Here, E and F refer to chain termination reactions; G refers to chain growth; H and J refer to branching and cross-linking reactions, respectively; and I refers to the formation of defect structures.

In the generalized form, the synthesis and chemical structure of polydiphenylenephthalide may be depicted as follows:

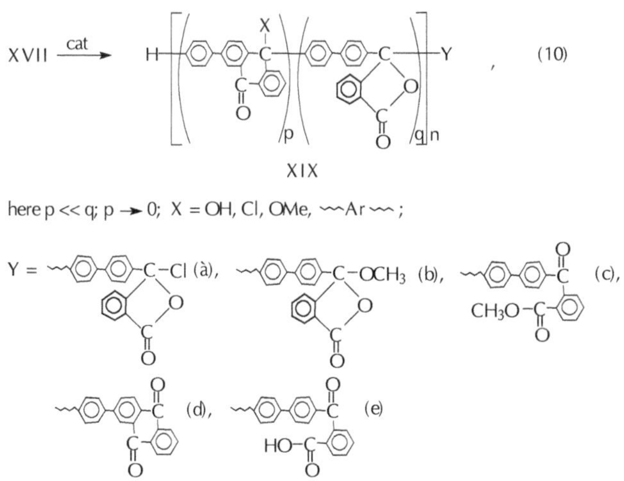

$$\text{XVII} \xrightarrow{\text{cat}} \quad (10)$$

XIX

here $p \ll q$; $p \rightarrow 0$; $X = OH, Cl, OMe, \sim\!\!\text{Ar}\!\!\sim$;

$Y = $ ... (à), ... (b), ... (c),

... (d), ... (e)

Yb, Yc, and X = OMe form from Ya and X = Cl upon precipitation of the polymer into methanol, and Ye and X = OH arise from Ya and X = Cl upon hydrolysis.

If the synthesis of polydiphenylenephthalide XVIII is carried out under optimum conditions, one may almost completely avoid the occurrence of branching and cross-linking reactions (the solution studies of this polymer and the data on its molecular mass, molecular mass distribution [35], and gelation [41, 42]). As evidenced by NMR measurements [29, 34], no branching takes place due to additional substitution of aromatic nuclei in diphenyl fragments located between phthalide groups. The formation of defect structures of the anthrone type can also be almost completely precluded [32] (indirect support for this statement is provided by the absence of gelation [41, 42] and branching [29, 34, 35] reactions). The ratio between the reactions of chain growth and chain termination is of primary importance for preparing high-molecular-mass polydiphenylenephthalide. The reaction of chain growth (intermolecular dehydrochlorinationreaction G in Scheme 9) is complicated by the following chain termination reactions: trivial hydrolysis (reaction E), leading to the formation of α-ketocarboxylic inactive end groups, and intramolecular dehydrochlorination (reaction F) that affords end anthraquinone inactive groups. A concurrent investigation into the influence of polycondensation conditions on the η_{red} value of the polymer and the reactions of chain growth and chain termination [32] allowed one to optimize the conditions of polycondensation; this allowed one not only to avoid the formation of defect structures but also to reduce the role of chain termination reactions. As a result, a linear high-molecular-mass polydiphenylenephthalide was synthesized [32–35]. Here, it is appropriate to mention that the notions of basic and side reactions are rather arbitrary and depend on the problem being solved. For example, for the synthesis of polyarylenephthalides, the formation of anthraquinone structures is an undesirable side reaction. However, when polyanthraquinones are prepared by the interaction of pyromellitic acid derivatives, including its dianhydride, with aromatic hydrocarbons [43], the formation of anthraquinone units is the basic reaction.

Later on, it was shown that the developed approach (Scheme 11) relying on the use of pseudoacid chlorides in polycondensation, governed by the electrophilic substitution mechanism, is of general importance and is applicable to the synthesis of not only polyarylenephthalides but of other polymers as well. Thus, the polycondensation of pseudoacid chlorides containing groups related to phthalidePhthalimidine [44, 45] and sulfophthalide groups [46] were accomplished as follows:

here R = —◯—◯— (a),   —◯—O—◯— (b),   ◯⌐◯ (c)
                                                          CH₂

Furthermore, intermediate products [47, 48] and monomers [49, 50], necessary for the synthesis of polyarylenephthalimidines XXI [47, 49] and polyarylenesulfophthalides XXIII [48, 50], were prepared. In the synthesis of polyarylenephthalimidines, only AlCl₃ served, in fact, as an efficient catalyst; this is apparently related to the formation of a stable complex between the catalyst and the nitrogen atom, with the optimum amounts of the catalyst being appreciably greater than that in the synthesis of polyarylenephthalides [44, 45]. In terms of specific features, the synthesis of polyarylenesulfophthalides is comparable with the preparation of polyarylenephthalides. Following the schemes presented above for the polycondensation of pseudoacid chlorides containing phthalimidine (XX, Scheme 11), phthalide (X, Scheme 5), and sulfophthalide (XXII, Scheme 11) groups, pseudoacid chlorides with essentially different capabilities for cyclochain isomerism may participate in polycondensation. Thus, in relation to phthalide-containing pseudoacid chlorides, which occupy an intermediate position with respect to isomerization ability, we are dealing with two extreme cases. Phthalimidine-containing pseudoacid chlorides are virtually devoid of this ability, as opposed to sulfophthalide-containing pseudoacid chlorides; the isomers of the latter compounds are easy to access even in the crystalline state. To synthesize polyarylenesulfophthalide, polycondensation of both acid chloride isomers was implemented [46].

(12)

XXII   XXIII   XXIV

The poorer outcome in the case of polycondensation of XXIV is probably accounted for by its worse purity compared to XXII.

In the general form, polycondensation, proceeding according to the electrophilic substitution mechanism with the use of pseudoacid chlorides as monomers, was described in a short comment [51] by the following scheme (for the polycondensation of monopseudoacid chlorides):

(13)

XXV   I

Here, X = –O–, N–R, Y = C=O, SO2, and H–R–H are aromatic hydrocarbons.

14.2.1 Properties

Polyarylenephthalides and their analogs synthesized by the electrophilic substitution reaction starting from pseudoacid chlorides exhibit a set of valuable properties: high thermal stability and good heat and chemical resistance, combined with excellent solubility in a variety of organic solvents.

Polyarylenephthalides containing polynuclear fragments without hinge atoms (XVIa, XVIb, XVIe) between phthalide groups are characterized by temperatures of the onset of softening (and glass transition temperatures) above 400°C. These polyarylenephthalides possess high thermal stability (the temperatures of the onset of degradation >440°C; in most cases, these values are in the range from 440 to 480°C) and are characterized by high coke numbers (up to 75–78). These polymers exhibit good solubility in many organic solvents, and strong transparent films may be cast from their solutions. The behavior of polydiphenylenephthalide was examined in more detail; the properties of this polymer will be considered in brief below as an example.

Polyarylenephthalides and their analogs containing diphenyl fragments between phthalide and related groups are most studied.

14.2.1.1   Polydiphenylenephthalide (PDPP)

The temperature of the onset of softening (and the glass transition temperature) for this polymer is nearly 420°C [26, 28, 45, 52] (the replacement of the diphenyl group by the terphenyl one leads to a rise in the temperature of the onset of softening by more than 50°C [52]), and the temperature of the onset of degradation is 3440°C (upon substitution of the diphenyl group by the terphenyl one, this temperature is nearly 480°C); the latter is determined using the most stringent test (based on the first signs of weight loss not higher than 1% rather than by the tangent method or from a 10% loss in weight) [26, 28, 45, 52]. The temperatures corresponding to the onset of degradation of PDPP in air and under inert atmosphere are almost similar [45]; when PDPP was heated up to 900°C under inert atmosphere or in the medium of its degradation products, a solid coke residue was formed in amount of 75–76% of the initial polymer weight (upon pyrolysis, the PDPP films acquired coloration but remained transparent and strong). The comprehensive study of thermal degradation (thermolysis, thermal oxidation, and hydrolysis) enabled one, not only to gain information valuable for practice, but also to provide insight into the processes of polymer chain degradation [53-61]. Thermal degradation studies may be conditionally divided into two directions: the study of low temperature degradation (in this temperature region, no loss in weight is observed) and the study of high temperature degradation (high temperature degradation was also examined using model compounds). When PDPP was heated in vacuum at rather low temperatures (300–400°C), virtually no gel fraction was formed; after heating at 425°C for 2 hr, the amount of the gel being formed was negligible (~1 wt %). Intense gelation took place only at 450°C. Upon heating in air to 325°C, no gel was formed; development of an appreciable amount of the gel was observed at 360–380°C. It was shown that the evolution of CO_2 is primarily related to the decomposition of the lactone cycle in the phthalide group and the decomposition of the end groups of type Y = XIXb, XIXc, and XIXe (Scheme (10)) results in the evolution of CO along with CO_2. The gelsol analysis and the [η] measurements of the sol fraction together with the data on gel swelling and the degree of its cross-linking made it possible to analyze processes leading to degradation, cross-linking, and branching. It was also demonstrated that the trace amounts of the catalysts affect the thermal stability of polydiphenylenephthalide. Additional information was obtained by comparing the data on the thermal degradation of PDPP with the analogos data reported in [62] for polyarylenephthalide with terphenyl moieties located between phthalide groups.

Based on the results from the thermal degradation of PDPP, it was inferred that there is the interrelationship between the structural transformations of phthalides, fluorene, fluorenone, anthraquinone, anthrone, anthracene, and triarylmethane units. This statement is analogous to our conclusion relying on the previously published data on the synthetic transformations of the polymers under consideration. It is suggested that processes that take place during the synthesis and thermal transformations of the above structures are interrelated and deserve generalization. The PDPP exhibits high chemical resistance [26, 45, 63] and excellent solubility in methylene chloride, chloroform, sym-tetrachloroethane, a phenol-sym-tetrachloroethane (1:3, wt/wt) mixture, o-chlorophenol, m-cresol, pyridine, N-methylpyrrolidone, N,N-dimethylacetamide, DMF, cyclohexanone, o- and m-dichlorobenzene, aniline, nitrobenzene, benzonitrile, and so on. Moreover, it is well soluble in epoxides (glycidyl phenyl ether). For example, at 100–130°C, this polymer dissolves in an ED-20 resin to form solution containing 30 g of this polymer in 100 g of the oligomer. The chemical structure of a radical situated between phthalide groups strongly affects the solubility of polyarylenephthalides. Thus, when the diphenyl radical is replaced by the diphenyl oxide moiety, the resulting polymer acquires solubility in sym-dichloroethane and THF while the substitution of the diphenyl radical by the terphenyl moiety results in worsening of solubility (the polymers lose solubility in methylene chloride, cyclohexanone, and DMF [52].

Valuable information about the thermodynamic quality of solvents was derived by studying the intrinsic viscosity of PDPP solutions in a wide variety of solvents [55].

Based on the analysis of the molecular mass characteristics of PDPP [35], it was concluded that this polymer lacks long-chain branching over the molecular mass range studied. This finding, in combination with the NMR data [29, 34], gives us grounds to state that the formation of branched structures during the synthesis of polydiphenylenephthalide may be avoided [35]. The $[\eta]$-M_w relation in a good solvent sym-tetrachloroethane and under θ-conditionswas estimated in [35]. Molecular masses were studied using various methods [35]: M_w was measured by light scattering and ultracentrifugation (the Archibald method), and M_n was determined by ebullioscopy, from the content of end groups, and by ultracentrifugation (the Lütje method). Of great significance is the testing of the mechanical properties of films prepared from individual fractions of PDPP; it was shown that this polymer can form films at $M_w > 2 \times 10^4$ and the film strength ceases to depend on molecular mass when M_w is $> 3.2 \times 10^4$.

Alongside the mechanical properties of PDPP-based films, relaxation transitions occurring in this polymer were investigated [64] (a b-transition at a temperature of 96°C and a glass transition temperature at 420°C). Of crucial importance was the examination of concentrated solutions, which revealed their specific features. At 25°C, PDPP solutions in chloroform and cyclohexanone, which contained more than 5–8 g of the polymer in 100 ml of the solvent, showed a tendency toward thermoreversible gelation; this phenomenon was observed for solutions containing 1030 g of the polymer in 100 ml of the solvent. An increase in the solution concentration and a reduction in temperature accelerated the process of gelation at T £ 25°C, the gelation time decreased from one month to several days, and, after heating at 40–60°C, the solution movable at room temperature was restored. The study of gels provided evidence that commonly known processes, such as syneresis and thixotropy, take place in them. When making articles from polymer solutions and during their subsequent drying, one should also take into account the stability of the polymer-solvent solvate complex. For example, in the course of PDPP film drying, even low-boiling solvents, such as methylene chloride and chloroform, can be removed only after long-term heating at 150°C. PDPP solutions were utilized not only for making fibers (primarily ultrathin), films, and coatings, but also as binding agents in the manufacture of strain gauges and as binding agents for their gluing [87]. Moreover, it was found that PDPP may be processed; not only from solutions but also from melt (monolithic plastics were prepared by molding). The PDPP possesses by very high chemical resistance at high temperature (200300°C) in concentrated solutions of hydrochloric acid, alkali, ammonia, and water; with this, the molecular mass of these polymers remains invariable (as estimated from the values of η_{red}). High chemical resistance of polyarylenephthalides opens wide possibilities for their chemical transformations without essential chemical destruction of macromolecule backbone.

One of the characteristic features of PDPP (and polyarylenephthalides in general) is their intense coloration in concentrated sulfuric acid. This phenomenon was discovered long ago [66, 67] and was initially used for quantitative determination of small amounts of the said polymers and visual and instrumental identification of the polymers falling into this group. Subsequently, this method was used for more delicate molecular and topological studies of PDPP and polyterphenylenephthalide [68-70], including transformations of their macromolecules during thermal degradation.

Another very important feature of polyarylenephthalides that was primarily investigated for the case of PDPP is essentially complex of properties, which permit to relate these polymers, first of all PDPP (it is most studied), to «smart» polymers [21, 22]. It is the sensitivity of its electrical, optical, and other properties to external actions (temperature, pressure, various kinds of radiation, electric and magnetic fields, etc.). This work was stimulated by earlier observations that some phthalides acquire coloration under pressure [71].

Polydiphenylene-N-phenylphtalimidine (XXIa) and polydiphenylenesulfophthalide (XXIIIa) are most studied and valuable analogies of polyarylenephthalides among polyarylenes related to polyarylenephthalides.

14.2.1.2 Polydiphenylene-N-phenylphtalimidine
This polymer ranks above PDPP in terms of thermal stability and heat resistance. However, this is true only for the polymer synthesized by the chemical transformation of PDPP in accordance with the following scheme:

$$XVIII \xrightarrow{H_2N-R} \qquad (14)$$

XXIà

here R = -C$_6$H$_5$, [45, 72, 73]

Polymer prepared by direct polycondensation cannot be completely purified from catalyst residues that markedly worsen its thermal stability. Polydiphenyl-ene-N-phenylphthalimidine, like polydiphenylenephthalide, exhibits high chemical resistance. Films cast from these polymers preserve excellent mechanical strength when exposed under pressure in concentrated solutions of hydrochloric acid, alkali, ammonia, and water at high temperatures (200300°C); with this, the molecular mass of these polymers remains invariable (as estimated from the values of η$_{red}$).

14.2.1.3 *Polydiphenylenesulfophthalide*

Polydiphenylenesulfophthalide is of interest not only as a thermally stable polymer (it strongly ranks below PDPP in this respect) but also as a latent polyelectrolyte capable of forming polymer salts. Unfortunately, publications in this area are only beginning to emerge [74, 75]. Spectral studies of colored salts based on polydiphenylenesulfophthalide and polyterphenylenesulfophthalide were described in [76–78].

14.3 CONCLUSION

Above described analysis shows validity of allocation among phthalide-containing polymers particular groups of polymers in which polymeric chain consist of alternation of phthalide groups and hydrocarbon structures without heteroatoms. At that polydiphenylenephthalide (XVIII) and poly(methylidene phthalide) (III) are deserved the most attention. In this case, the most content of phthalide group in polymer chain is achieved. As a result, influence of phthalide groups on the polymers properties is developed in the most extent for these polymers. On one hand, they possess by high heat and chemostability, good solubility in many solvents, which are general properties for these two polymers and on the other hand, they have differences in the properties. For instance, differences in thermostability, high thermostability for XVIII, and relatively moderate thermostability for III because of their disposition to depolarization were revealed. Some differences in the solubility and number of other properties for these polymers were also discovered. Taking into account these features, future trends of practical application of these polymers may be correctly determined. In the case of XVIII number of unusual properties was revealed: first of all effect of electron switching in result of external influences (temperature, pressure, electrical, and magnetic fields), gigantic magnetoresistance, and others. In this direction, investigation of polymer III is necessary, since poly(methylidene phthalide) and their copolymers may possess by unusual properties like polydiphenylenephthalide [20] and some other phthalide-containing polymers [7, 21]. Sharp increasing of these specific properties in the case of phthalide-containing copolyaryleneetherketones [21] permits to consider that investigation of binary and triple copolymers of methylidene phthalide is promising.

KEYWORDS

- **Depolymerization**
- **Intramolecular dehydrochlorination**
- **Methylidene phthalide**
- **Polyarylenephthalides**
- **Polycondensation**
- **Poly(diphenylene phthalide)**
- **Thermoreversible gelation**

ACKNOWLEDGMENT

This work has been financially supported by the Russian Foundation for Basic Research (grant 10-03-00648-a), the Presidium of RAS (grant 7P).

REFERENCES

1. Vinogradova, S. V. and Vygodskii, Ya. S. Cardo Polymers. *Usp. Khim.*, **42**, 1225 (1973).
2. Korshak, V. V., Vinogradova, S. V., and Vygodskii, Ya. S. Cardo Polymers. *J. Macromol. Sci., Rev. Macromol. Chem.*, **11**, 45 (1974).
3. Vygodskii, Ya. S. and Vinogradova, S. V. Peculiarities of Synthesis and Properties of Cardo Polymers. In *Chemistry and Technology of High Polymers* (VINITI Akad. Nauk SSSR, Moscow), 7, p. 14 (1975).
4. Vinogradova, S. V., Vasnev, V. A., and Vygodskii, Ya. S. Cardo Polyheteroarylenes. Synthesis, Properties, Peculiarities. *Usp. Khim.*, **65**, 266 (1996).
5. Salazkin, S. N. and Chelidze, G. Sh. Synthesis of Methylidene Phthalide, Its Polymerization, and Properties of Poly(methylidene phthalide). *Polymer Science, Ser. C*, **51**(1), 126 (2009) [*Vysokomol. Soedin., Ser.* C **51**, 1386 (2009)]
6. Salazkin, S. N. Aromatic Polymers Based on Pseudoacid Chlorides.*Polymer Science, Ser. B* **46**, 203 (2004) [*Vysokomol. Soedin., Ser. B* **46**, 1244 (2004)].
7. Salazkin, S. N., Shaposhnikova, V. V., Machulenko, L. N. et al. Synthesis of Polyarylenephthalides Prospective as Smart Polymers *Polymer Science, Ser. A* **50**, 243 (2008) [*Vysokomol. Soedin., Ser. A* **50**, 399 (2008)].
8. Gabriel, S. Ueber die constitution der phtalylessigssäure. *Ber.*, **17**, 2521 (1884).
9. Coower, H. W., Shearer, N. H., and Dickey, J. B. Polymers of 3-methylenephthalide. US Patent No. 2618627 (1952).
10. Vinogradova, S. V., Salazkin, S. N., Korshak, V. V. et al. Synthesis and Behavior of Polymethylidenphthalide.*Vysokomol. Soedin., Ser. A* **12**, 205 (1970).
11. Rode, V. V., Zhuravleva, I. V., Gamza-Zade, A. I. et al. Thermostability of Polymethylidenphthalide.*Izv. Akad. Nauk SSSR, Ser. Khim.*, (4), 926 (1970).
12. Korshak, V. V., Salazkin, S. N., Vinogradova, S. V. et al. Concerning The Influence of Large Rings in The Side on the Thermostability of Hydrocarbon Polymers. *Vysokomol. Soedin., Ser. B* **13**, 150 (1971).
13. Vinogradova, S. V., Salazkin, S. N., Chelidze, G. Sh. et al. Co-polymers of methylenephthalide. *Plast. Massy*, (8), 10 (1971).
14. Mowry, D. T. and Mills, Ch. L. Styren-alkylidene phthalide copolymers. US Patent No. 2489972 (1948).

15. Yale, H. L. O-Acetobenzoic Acid, its Preparation and Lactonization. A Novel Application of the Doebner Synthesis. *J. Am. Chem. Soc.*, **69**, 1547 (1947).
16. Kariyone, T. and Shimizu, S. Synthesis of Alkylidenephthalides and Their Odor.J. Pharm. *Soc. Jpn.*, **73**, 336 (1953).
17. Vinogradova, S. V., Korshak, V. V., Salazkin, S. N., and Chelidze, G. Sh. Synthesis of Phthalide-nacetic Acid and methylidenphthalide. *Zh. Prikl. Khim.* (*Leningrad*) **44**, 1389 (1971).
18. Ito, H. and Ueda, M. Polymerization of Exo-methylene Styrenic Monomers: effect of a,a-cyclization on reactivity and stereoregularity.*Polymer Preprints*, **32**(1), 428 (1991).
19. Chelidze, G. Sh. and Salazkin, S. N. Influence of Water on Polymerization of Polymethylidene phthalide.*Izv. Akad. Nauk SSSR, Ser. Khim.*, (8), 1882 (1972).
20. Lachinov, A. N. and Vorob'eva, N. V. Electronics of Thin Wideband Polymer Layers.Usp. *Fiz. Nauk*, **176**, 1249 (2006).
21. Salazkin, S. N. and Shaposhnikova, V.V. Synthesis of Phthalide-containing Polymers Having Great Potential for Creation of Functional Materials of Various Purposes. *Nanotekhnologii* (*nauka i proizvodstvo*), **3**(4), 3 (2009).
22. Korshak, V. V., Vinogradova, S. V., Chelidze, G. Sh. et al. Synthesis and Investigation of Poly-methylidene phthalimidines.*Vysokomol. Soedin., Ser. A*, **14**, 1496 (1972).
23. Vinogradova, S. V., Salazkin, S. N., Korshak, V. V., et al. Synthesis of methylidene phthalimidine. *Zh. Prikl. Khim.*, (*Leningrad*) **45**, 1813 (1972).
24. Rafikov, S. R., Tolstikov, G. A., Salazkin, S. N., and Zolotukhin, M. G. Polyheteroarylenes for Preparation of Thermostable Materials and Method of Their Preparation.USSR Invertor's Certificate No. 704114. *Byull. Izobret.*, (27) (1981).
25. Rafikov, S. R., Tolstikov, G. A., Salazkin, S. N., and Zolotukhin, M. G. Polyheteroarylenes for Preparation of Thermostable Materials and Method of Their Preparation. USSR Invertor's Certificate No. 734989. *Byull. Izobret.*, (20) (1981).
26. Salazkin, S. N., Rafikov, S. R., Tolstikov, G. A. et al. PolyarylidesNovel Aromatic Polymers. Available from VINITI, No. 3905 (Moscow, 1980).
27. Salazkin, S. N., Rafikov, S. R., Tolstikov, G. A., and Zolotukhin, M. G. Synthesis of Aromatic Polymers by Polycondensation Through Electrophylic Substitution Reaction With Participation of Pseudoacid Chlorides. Available from VINITI, No. 4310 (Moscow, 1980).
28. Salazkin, S. N., Rafikov, S. R., Tolstikov, G. A., and Zolotukhin, M. G. A New Way of Synthesis of Aromatic Polymers.*Dokl. Akad. Nauk SSSR*, **262**, 355 (1982).
29. Salazkin, S. N. and Rafikov, S. R. Use of Electrophylic Substitution Reaction for Synthesis of Polyheteroarylenes. *Izv. Akad. Nauk KazSSR, Ser. Khim.*, (5), 27 (1981).
30. Zolotukhin, M. G. Egorov, A. E., Sedova, E. A., et al. Dichloroanhydrides of Bis-(Ortho-Ketocarbonic Acids) as Perspective Monomers for Synthesis of Novel Polymers. *Dokl. Akad. Nauk SSSR*, **311**, 127 (1990).
31. Zolotukhin, M. G., Sedova, E. A., Salazkin, S. N., and Rafikov, S. R. Synthesis and structure of some Chloroanhydrides of Ortho-Ketocarbonic Acids. Available from VINITI, No. 6009 (Moscow, 1983).
32. Kovardakov, V. A., Zolotukhin, M. G., Salazkin, S. N., and Rafikov, S. R. Polydiphenylene-phthalide. Polycondensation of para-(3-Chloro-3-Phthalidyl)Diphenyl: Selectivity and Ration of Chain Growth Reactions and Chain Termination Reactions. Available from VINITI, No. 5606 (Moscow, 1983).
33. Zolotukhin, M. G., Kovardakov, V. A., Salazkin, S. N., and Rafikov, S. R. Regularities of Synthesis of Polydiphenylene Phthalide by Polycondensation of para-(3-Chloro-3-Phthalidyl)Diphenyl. *Vysokomol. Soedin., Ser., A* **26**, 1212 (1984).
34. Zolotukhin, M. G., *Panasenko*, A. A., Sultanova, V. S., et al. NMR-Study of Poly(phthalidylidenearylene)s. *Makromol. Chem.*, **186**, 1747 (1985).
35. Salazkin, S. N., Zolotukhin, M. G., Kovardakov, V. A., et al. Molecular Mass Characteristics of Poly(diphenylene Phthalide). *Vysokomol. Soedin., Ser. A*, **29**, 1431 (1987).

36. Rafikov, S. R., Salazkin, S. N., Zolotukhin, M. G., and Kovardakov, V.A. Method of Preparation of beta-Phenylanthraquinone. USSR Invertor's Certificate No. 1020422. *Byull. Izobret.*, (20) (1983).
37. Kovardakov, V. A., Zolotukhin, M. G., Salazkin, S. N., and Rafikov, S. R. Synthesis of Phenyl-anthraquinone by Thermal Intramolecular Dehydrochlorinationof para-(3-Chloro-3-Phthalidyl) Diphenyl. *Izv. Akad. Nauk SSSR, Ser. Khim.*, (4), 941 (1983).
38. Kovardakov, V. A., Zolotukhin, M. G., Kapina, A. P. et al. *Investigation of Transformations of para-(3-Chloro-3-Phthalidyl)Diphenyl by Thermal Influence.* Available from VINITI, No. 5089 (Moscow, 1982).
39. Kovardakov, V. A., Nikiforova, G. I., Kapina, A. P. et al. Polydiphenylenephthalide: *Quantitative Analysis of End Groups and Side Structures in Macromolecules.* Available from VINITI, No. 2773 (Moscow, 1983).
40. Kovardakov, V. A., Sokol'skaya, O. V., Salazkin, S. N., and Rafikov, S. R. Polydiphenylene-phthalide. Investigation of Polycondensation of para-(3-Chloro-3-Phthalidyl)Diphenyl: *Proce-dure of Isolation and Treatment of Polymer Simplifying Correct Quantitative Control of End Groups Content.* Available from VINITI, No. 4892 (Moscow, 1983).
41. Zolotukhin, M. G., Skirda, V. D., Sundukov, V. I. et al. Reasons of Gelation in Synthesis of Polyarylenephthalides. *Vysokomol. Soedin. Ser.*, B **29**, 378 (1987).
42. Zolotukhin, M. G., Skirda, V. S., Sedova, E. A. et al. Poly(phthalidylidenarylene)s. Gelation in The Homopolycondensation of 3-Aryl-3-chlorophthalides. *Makromol. Chem.*, **194**, 543 (1993).
43. Saltybaev, D. K., Zhubanov, B. A., and Pivovarova, L.V. Ladder and Partially-Ladder Polyqui-nones. *Vysokomol. Soedin., Ser. A*, **21**, 734 (1979).
44. Rafikov, S. R., Salazkin, S. N. Shumanskii, M. E., and Akhmetzyanov, Sh. S. Method of Prepara-tion of Polyheteroarylenes with Phthalimidine groups. USSR Invertor's Certificate No. 966093. *Byull. Izobret.*, 38 (1982).
45. Salazkin, S. N., Belen'kaya, S. K., Zemskova, Z. G. et al. Polyarylenephthalimidines. *Dokl. Akad. Nauk.*, **357**, 68 (1997).
46. Zolotukhin, M. G., Akhmetzyanov, Sh. S., Lachinov, A. N. et al. Poly(arylenesulfophthalide)s. *Dokl. Akad. Nauk SSSR*, **312**, 1134 (1990).
47. Rafikov, S. R., Salazkin, S. N., Shumanskii, M. E., and Akhmetzyanov, Sh. S. Derivatives of Phthalimidines as Intermediate Compounds for Synthesis of Monomers Used in Preparation of Thermostable Polymers. USSR Invertor's Certificate No. 1012575. *Byull. Izobret.*, (1) (1984).
48. Akhmetzyanov, Sh. S., Tolstikov, G. A., Lachinov, A. N. et al.Derivatives of ortho-Sulfobenzoic Acid as Intermediate Compounds for Synthesis of Monomers Used in Preparation of Soluble Polymers Possessing by Electroconductivity after Dopation by Iodine. USSR Invertor's Certifi-cate No. 1467054. *Byull. Izobret.*, (11) (1989).
49. Rafikov, S. R., Salazkin, S. N., Shumanskii, M. E., and Akhmetzyanov, Sh. S. Derivatives of Phthalimidines as Monomers for Synthesis of Thermostable Polymers.USSR Invertor's Certifi-cate No. 1015610. *Byull. Izobret.*, (47) (1983).
50. Akhmetzyanov, Sh. S., Tolstikov, G. A., Lachinov, A. N. et al. Derivatives of ortho-Sulfobenzoic Acid as Monomers for Synthesis of Soluble Polymers Possessing by Electroconductivity after Dopation by Iodine. USSR Invertor's Certificate No. 1467060. *Byull. Izobret.*, (11) (1989).
51. Salazkin, S. N. Synthesis of Poly(diphenylene phthalide) by Polycondensation of p-(3-Chloro-3-Phthalidyl)diphenyl - Comment.*Vysokomol. Soedin. Ser. A*, **41**, 1989 (1999) [*Polymer Science, Ser. A*, **41**, 1273 (1999)].
52. Sigaeva, N.N., Zolotukhin, M. G., Kozlov, V. G. et al. Molecular and Hydrodynamic Charac-teristics of Poly(terphenylene phthalide). *Vysokomol. Soedin., Ser. B*, **37**, 2066 (1995) [*Polymer Science, Ser. B*, **37**, 557 (1995)].
53. Kraikin, V. A., Laktionov, V. M., Zolotukhin, M. G. et al. High temperature Degradation of Polyarylene Phthalides.Available from VINITI, No. 5021 (Moscow, 1985).
54. Kraikin, V. A., Salazkin, S. N., Komissarov, V. D., et al. Thermodegradation of Polyarylene Phthalides. *Vysokomol. Soedin. Ser. B*, **28**, 264 (1986).

55. Kraikin, V. A., Laktionov, V. M., Zolotukhin, M. G. et al. Low temperature Degradation of Polyarylene Phthalides.Available from VINITI, No. 3081 V (Moscow, 1991).
56. Kraikin, V. A., Kovardakov, V. A., Panasenko, A. A., et al., Pyrolysis of Poly(Arylene Phthalides). *Vysokomol. Soedin. Ser. A*, **34**, 28 (1992) [*Polymer Science* **34**, 478 (1992)].
57. Kraikin, V. A., Laktionov, V. M., Zolotukhin, M. G., et al. Degradation of Poly(arylenephthalide) s in Air and Under Vacuum.*Vysokomol. Soedin. Ser., A* **40**, 1493 (1998) [*Polymer Science Ser., A* **40**, 934 (1998)].
58. Kraikin, V. A., Belen'kaya, S. K., Sedova, E. A., et al. Influence of Polyheteroarylenes Preparation Conditions on Their Thermostability. *Bashk. Khim. Zh.*, **6**, 39 (1999).
59. Kraikin, V.A., Sedova, E. A., Musina, Z. N., and Salazkin, S. N. Influence of Catalyst Residue on Thermostability of Polydiphenylenephthalimidine.*Vestn. Bashk. Gos. Univ.*, (1), 45 (2001).
60. Kraikin, V. A., Kovardakov, V. A., and Salazkin, S. N. Thermal Transformations of Poly(diphenylenephthalide) and Its Low-molecular-mass Models. *Vysokomol. Soedin., Ser. A* **43**, 1399 (2001) [*Polymer Science, Ser. A*, **43**, 885 (2001)].
61. Kraikin, V.A., Kuznetsov, S. I., Laktionov, V. N., and Salazkin, S.N.. Thermal Oxidation and Thermal Hydrolysis of Poly(arylenephthalides). *Vysokomol. Soedin., Ser. A*, **44**, 834 (2002) [*Polymer Science, Ser. A*, **44**, 518 (2002)].
62. Kraikin, V. A., Egorov, A. E., Musina, Z. N. et al. Thermal Properties and Degradation Behavior of Poly(arylenephthalides) Based on 4,4'-Bis(3-Chloro-3-Phthalidyl)terphenyl. *Vysokomol. Soedin., Ser. A* **42**, 1574 (2000) [*Polymer Science, Ser. A* **42**, 1046 (2000)].
63. Rafikov, S. R., Salazkin, S. N., and Zolotukhin, M. G. Resistance of Poly(arylenephthalides) to Action of Aggressive Medium. *Plast. Massy*, (10), 56 (1986).
64. Nikol'skii, O. G., Askadskii, A. A., Salazkin, S. N., and Slonimskii, G. L. Investigation of Relaxation Transitions in Aromatic Heat-resistant Polymers. *Mekh. Kompoz. Mater. (Rizh. Politekh. Inst.)*, (6), 963 (1983).
65. Ser'eznov, A. N., Tsareva, G. A., Oboturova, N. A. et al. Binder for Making and Gluing of High temperature Resistive-strain Sensor. USSR Invertor's Certificate No. 757842. *Byull. Izobret.*, (31) (1980).
66. Kraikin, V. A., Zolotukhin, M. G., Salazkin, S. N., and Rafikov, S. R. Method of Quantitative Determination of Polyarylenephthalides. USSR Invertor's Certificate No. 1065741. *Byull. Izobret.*, (1) (1984).
67. Kraikin, V. A., Zolotukhin, M. G., Salazkin, S.N., and Rafikov, S. R. Qualitative and Quantitative-Determinations of Polyarylene Phthalides Based on Their Capacity to Form the Intensively Colored Solutions in Concentrated Sulfuric-Acid. *Vysokomol. Soedin. Ser. A*, **27**, 422 (1985).
68. Kraikin, V.A., Egorov, A. E., Puzin, Yu. I. et al. The Spectra Parameters of The Poly(terphenylenephthalide) Sulfuric Acid Solutions and The Polymer Chain Length. *Dokl. Akad. Nauk*, **367**, 509 (1999).
69. Kraikin, V. A., Musina, Z. N., Egorov, A. E. et al. Conjugation in Chromophore Groups of Low- and High-molecular-weight Arylenephthalides and an Equation of Main Absorption Wavelength. *Dokl. Akad. Nauk.*, **372**, 66 (2000).
70. Kraikin, V. A. Musina, Z. N., Sedova, E. A. et al. New Approach to The Study of The Topology of Branched Phthalide-containing Polyheteroarylenes. *Dokl. Akad. Nauk*, **373**, 635 (2000).
71. Petrov, A.A., Gonikberg, G.G., Aneli, Dzh. N. et al. Behavior of Substituted Diphenylphthalides and Related Lactams Under High Pressures and Shearing Stresses. *Izv. Akad. Nauk SSSR, Ser. Khim.*, (2), 279 (1968).
72. Rafikov, S. R., Salazkin, S. N. Zemskova, Z. G. and Belen'kaya, S. K. Polyheteroarylenes for Making of Thermostable Materials and Method of their preparation, USSR Invertor's Certificate No. 860484. *Byull. Izobret.*, (13) (1982).
73. Salazkin, S. N. Belen'kaya, S. K. Kraikin, V. A. et al. Synthesis and Some Properties of Poly(diphenylene-N-arylphthalimidines). Available from VINITI, No. 1997 (Moscow, 1985).
74. Vasil'ev, V. G., Nikiforova, G. G., Rogovina, L. Z., et al. Poly(diphenylene sulfophthalide) Derivatives with Polyelectrolyte and Specific Optical Properties. *Dokl. Akad. Nauk*, **382**, 649 (2002).

75. Rogovina, L. Z., Vasil'ev, V. G., Nikiforova, G. G., et al. Poly(diphenylenesulfophthalide) and the related alkali-metal salts. *Vysokomol. Soedin., Ser. A*, **44**, 1302 (2002) [*Polymer Science, Ser. A*, **44**, 817 (2002)].
76. Shishlov, N. M., Akhmetzyanov, Sh. S., and Khrustaleva, V. N. Color Reactions of Polyarylene-sulfophthalides in AnilineCyclohexanone Mixtures in Air. *Izv. Ross. Akad. Nauk, Ser. Khim.*, (2), 389 (1997).
77. Shishlov, N. M., Khrustaleva, V. N., Akhmetzyanov, Sh. S. et al. Formation of Color Centers and Paramagnetic Species by Alkaline Hydrolysis of Polydiphenylenesulfophthalide. *Izv. Ross. Akad. Nauk, Ser. Khim.*, (2), 295 (2000).
78. Shishlov, N. M., Khrustaleva, V. N., Akhmetzyanov, Sh. S. et al. Formation of Color Centers and Polyradicals on Reduction of Polydiphenylenesulfophthalide by Metallic Lithium. *Dokl. Akad. Nauk*, **374**, 206 (2000).

15 Prediction of Reinforcement Degree for Nanocomposites Polymer/Carbon Nanotubes

Z. M. Zhirikova, G. V. Kozlov, and V. Z. Aloev

CONTENTS

15.1 INTRODUCTION

The quantitative scaling model for nanocomposites polymer/carbon nanotubes reinforcement degree prediction has been offered. A radius of ring-like structures, formed by nanotubes in polymer matrix, usage for reduction factor determination is a specific feature of this model. At nanocomposites reinforcement degree calculation, the offered model does not take into consideration a nanofiller elasticity modulus.

Carbon nanotubes (nanofibers) have two specific features. First, they are normally supplied as little bundles of entangled nanotubes, dispersion of which is difficult enough [1]. Second, carbon nanotubes possess very high longitudinal stiffness (high elasticity modulus) while their flexural rigidity is very low for geometric reasons [2]. This property defines distortion (departure from rectilinearity) of the initial nanotubes geometry, namely, their rolling up in ring-like structures [2]. However, at small concentrations, as those used in the present work, the formation of aggregates is scarcely probable [2]. The purpose of the present paper is the development of a quantitative model including both the longitudinal stiffness and the flexibility as parameters for predicting the effectiveness of nanotubes reinforcement.

15.2 EXPERIMENTAL

Polypropylene (PP) "Kaplen" commercially available, having average-weight molecular weight of $\sim (23) \times 10^5$ and polydispersity index of 4.5, was used as a matrix polymer. Two types of carbon nanotubes were used as nanofiller. Nanotubes of mark "Taunite" (CNT) have an external diameter of 2070 nm, an internal diameter 510 nm, and length of 2 mcm and more. Besides, multiwalled nanofibers (CNF), having layers number of 2030, a diameter of 2030 nm and length of 2 microns and more, have been used. The mass contents of carbon nanotubes of both types were changed in the range of 0.153.0 mass %.

15.3   DISCUSSION AND RESULTS

As it is noted above, carbon nanotubes rectilinearity distortion or their bending (rolling up) is due to their very high flexibility, and can be one of the main model parameters. The distortion degree can be estimated with the help of forming ring-like structures radius R_{CNT} [3] as follows:

$$R_{CNT} = \left(\frac{\pi L_{CNT} D_{CNT}^2}{32 j_n} \right)^{1/3} \tag{1}$$

where L_{CNT} and D_{CNT} are length and diameter of carbon nanotubes. For CNT and CNF the value L_{CNT} was fixed to 2 μm.

The scaling model, already presented in chapter [4], can be used for quantitative treatment of carbon nanotubes rolling up processes in order to evaluate its influence on nanocomposites stiffness.

The gist of this model consists of the introduction of reduction factor α^{n-1}, connecting filler mass concentrations W_n in two equivalent composites A and B as follows:

$$W_n^B = \alpha^{n-3} W_n^A \tag{2}$$

where W_n^A and W_n^B is the mass content in composites A and B, respectively, a is the aspect ratio of aggregates or particles, n is the parameter characterizing the shape of the filler particles, assumed equal to 1 for short fibers, 2, for disk like (flaky) particles and 3, for spherical particles.

Within the frameworks of the model, the relationship between the filler particles size D_p and composite stiffness E_c [4] is defined as follows:

$$E_c \left(\theta, W_n, D_p \right) = E_c \left(\theta, \alpha^{n-3}, W_n \alpha D_p \right) \tag{3}$$

where q is the parameter reflecting the particles size distribution.

The model [4] assumes that the elasticity modulus depends on the reduced filler contents, corrected on the basis of particle shape (n) and size (D_p).

Carbon nanotubes behavior as nanofiller in polymer matrix, requires a reduction factor a calculated as follows:

$$\alpha = \frac{R_{CNT}^{max}}{R_{CNT}} \tag{4}$$

where R_{CNT}^{max} and R_{CNT} are maximum and current radii of ring-like structures, which form carbon nanotubes.

Calculation according to the Equation (1) has shown that the value $R_{CNT}^{max} = 1.33$ microns for CNT and $R_{CNT}^{max} = 0.71$ microns for CNF. In CNT nanotubes, varying mass content from 0.25 to 3.0 mass % results in a R_{CNT} reduction from 1.33 to 0.58 microns, while for CNF nanotubes with the mass content varying from 0.15 to 3.0% R_{CNT} reduces from 0.71 down to 0.26 mcm. The exponent n in the Equations (2) and (3) was fixed equal to 1. Hence, the final formula for reduction factor has the following form:

$$\alpha^{n-3} = \left(\frac{R_{CNT}^{max}}{R_{CNT}}\right)^{-2} \qquad (5)$$

In Figure 1, the reinforcement effectiveness, E_n/E_m is plotted as function on the combined parameter $\alpha^{n-3}W_n$ for nanocomposites PP/CNT and PP/CNF. As one can see, the reinforcement effectiveness is described by a straight line that by best fitting the data can be expressed analytically as follows:

$$\frac{E_n}{E_m} = 1 + 0.65\alpha^{-2}W_n \qquad (6)$$

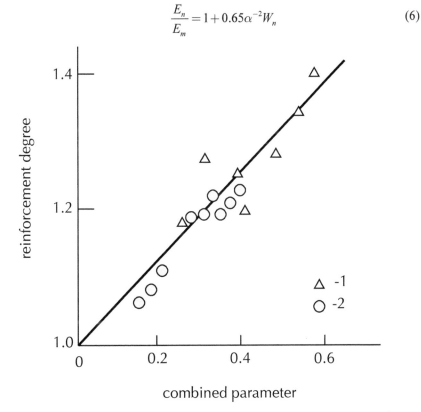

FIGURE 1   The dependence of reinforcement degree E_n/E_m on combined parameter $\alpha^2 W_n$ for nanocomposites PP/CNT (1) and PP/CNF (2).

KEYWORDS

- **Carbon nanotubes**
- **Nanocomposite**
- **Reinforcement degree**
- **Ring-like structures**
- **Scaling model**

REFERENCES

1. Rakov, E. G. *Uspekhi Khimii*, **76**(1), 3–19 (2007).
2. Sow, C. H., Lim, K. Y., Cheong, F. C., and Saurakhiya, M. *Current Research on Nanotechnology*, **1**(2), 125–154 (2007).
3. Yang, Y., D'Amore, A., Di, Y., Nicolais, L., Li, B. *Journal of Applied Polymer Science*, **59**(7), 1159–1166 (1996).
4. Bridge, B. *J. Mater. Sci. Lett.*, **8**(2), 102–103 (1989).
5. Koiwai, A., Kawasumi, M., Hyodo, S., Motohiro, T., Noda, S., and Kamigaito, O. *Mater. of Intern. Symp.* "Benibana", Yamagata, Japan, 105–110 (1990).
6. Netti, P., D'Amore, A., Ronca, D., Ambrosio, L., and Nicolais, L. *Journal of Materials Science: Materials in Medicine*, **7**(9), 525–530 (1996).
7. Caprino, G., D'Amore, A. and Facciolo, F. *Journal of Composite Materials*, **32**(12), 1203–1220 (1998).

16 Application of Polymers in Construction Technology (PART I)

Effects of elevated temperature on polypropylene fiber reinforced mortar

A. K. Haghi and G. E. Zaikov

CONTENTS

16.1 INTRODUCTION

Municipal solid waste (MSW) management is a significant contributor of greenhouse gas (GHG) emissions and especially the disposal of waste in landfills generates methane (CH_4) that has high global warming potential. Waste management activities and especially disposal of waste in landfills contribute to global GHG emissions approximately by 4%. The most common methods used for MSW, besides landfilling, include composting, recycling, mechanical-biological treatment (MBT) and waste-to-energy (WTE). The European waste policy force diversion from landfill and WTE is a waste management option that could provide diversion from landfill and at the same time save a significant amount of GHG emissions, since it recovers energy from waste which usually replaces an equivalent amount of energy generated from fossil fuels. However, disposal of MSW in sanitary landfills is still the main waste management method in many countries, both in the EU and internationally, although diversion from landfilling is generally promoted and the perspectives of new waste treatment technologies also evaluated. Thus, there are quite a few "developed" countries which are really still in a developing stage in terms of sustainable MSW management and the balanced integration of

WTE in their overall system. In this chapter, we will show how to convert the recycled wastes into wealth with particular application in construction industries.

16.2 EXPERIMENTAL

Materials used included ordinary Portland cement type 1, standard sand, silica fume, glass with tow particle size, rice husk ash (RH), tap water, and finally fibrillated polypropylene fibers.

The fibers included in this study were monofilament fibers obtained from industrial recycled raw materials that were cut in factory to 6 mm length. Properties of waste Polypropylene fibers are reported in Table 1 and Figure 1.

TABLE 1   Properties of polypropylen fibers.

Property	Polypropylene
Unit weight [g/cm3]	0.9 – 0.91
Reaction with water	Hydrophobic
Tensile strength [ksi]	4.5 – 6.0
Elongation at break [%]	100 – 600
Melting point [°C]	175
Thermal conductivity [W/m/K]	0.12

FIGURE 1 Polypropylene fiber used in this study.

In accordance to ASTM C618, the glass satisfies the basic chemical requirements for a pozzolanic material especially clean glass. To satisfy the physical requirements for fineness, the glass has to be grounded to pass a 45 μm sieve.

Also the silica fume and RH contain 91.1 and 92.1% SiO_2 with average size of 7.38 μm and 15.83 μm respectively were used. The chemical compositions of all pozzolanic materials containing the reused glass, silica fume and RH were analyzed using an x-ray microprobe analyzer and listed in Table 2.

TABLE 2 Chemical composition of materials.

	Content (%)		
Oxide	Glass C	Silica fume	Rice husk ash
SiO_2	72.5	91.1	92.1
Al_2O_3	1.06	1.55	0.41
Fe_2O_3	0.36	2	0.21
CaO	8	2.24	0.41
MgO	4.18	0.6	0.45
Na_2O	13.1	–	0.08
K_2O	0.26	–	2.31
CL	0.05	–	–
SO_3	0.18	0.45	–
L.O.I	–	2.1	–

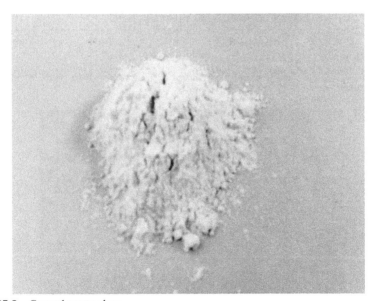

FIGURE 2 Ground waste glass.

To obtain this aim recycled windows clean glass was crushed and grinded in laboratory and sieved the ground glass to the desired particle size (Figure 2). To study the particle size effect, two different ground glasses were used, namely:

- *Type I*: Ground glass having particles passing a #80 sieve (180 μm).
- *Type II*: Ground glass having particles passing a #200 sieve (75 μm).

In addition the particle size distribution for two types of ground glass, silica fume, RH, and ordinary Portland cement were analyzed by laser particle size set, have shown in Figure 3. As it can be seen in Figure 3 silica fume has the finest particle size. According to ASTM C618, fine ground glasses under 45 μm qualify as a pozzolan due to the fine particle size. Moreover, glass type I and II respectively have 42% and 70% fine particles smaller than 45 μm that causes pozzolanic behavior. The SEM particle shape of tow kind of glasses is illustrated in Figure 4.

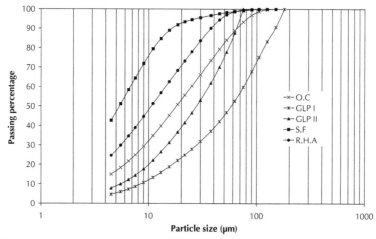

FIGURE 3 Particle size distribution of ground waste glass type I, II, silica fume, RH and ordinary cement.

FIGURE 4 *(Continued)*

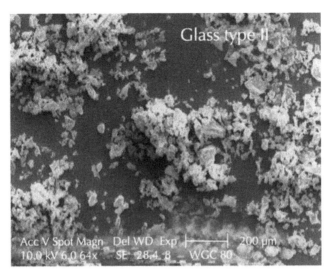

FIGURE 4   Particle shape of ground waste glass type I, type II.

16.3   TEST PROGRAM

For the present study, twenty batches were prepared. Control mixes was designed containing standard sand at a ratio of 2.25:1 to the cement in matrix. A partial replacements of cement with pozzolans include ground waste glass (GI, GII), silica fume (SF), and RH were used to examine the effects of pozzolanic materials on mechanical properties of PP reinforced mortars at high temperatures. The amount of pozzolans which replaced were 10% by weight of cement which is the rang that is most often used.

Meanwhile, polypropylene fibers were used as addition by volume fraction of specimens. The reinforced mixtures contained PP fiber with three designated fiber contents of 0.5, 1, and 1.5% by total volume.

In the plain batches without any fibers, water to cementitious ratio of 0.47 was used whereas in modified mixes (with different amount of PP fibers) it changed to 0.6 due to water absorption of fibers. The mix proportions of mortars are given in Table 3.

The strength criteria of mortar specimens and impacts of polypropylene fibers on characteristics of them were evaluated at the age of 60 days.

In our laboratory, the test program mix conducted as follows:

(1) The fibers were placed in the mixer.
(2) Three quarters of the water was added to the fibers while the mixer was running at 60 rpm: mixing continues for 1 min.
(3) The cement was gradually the cement to mix with the water.
(4) The sand and remaining water were added, and the mixer was allowed to run for another 2 min.

After mixing, the samples were casted into the forms 50 × 50 × 50 mm for compressive strength and 50 × 50 × 200 mm for flexural strength tests. All the moulds were coated with mineral oil to facilitate demoulding. The samples were placed in two lay-

TABLE 3 Mixter properties.

Batch No	sand/c	w/c	Content (by weight)					PP fibers (by volume)
			O.C	GI	GII	SF	RH	
1	2.25	0.47	100	–	–	–	–	0
2	2.25	0.47	90	10	–	–	–	0
3	2.25	0.47	90	–	10	–	–	0
4	2.25	0.47	90	–	–	10	–	0
5	2.25	0.47	90	–	–	–	10	0
6	2.25	0.6	100	–	–	–	–	0.5
7	2.25	0.6	90	10	–	–	–	0.5
8	2.25	0.6	90	–	10	–	–	0.5
9	2.25	0.6	90	–	–	10	–	0.5
10	2.25	0.6	90	–	–	–	10	0.5

batch No	sand/c	w/c	Content (by weight)					PP fibers (by volume)
			O.C	GI	GII	SF	RH	
11	2.25	0.6	100	–	–	–	–	1
12	2.25	0.6	90	10	–	–	–	1
13	2.25	0.6	90	–	10	–	–	1
14	2.25	0.6	90	–	–	10	–	1
15	2.25	0.6	90	–	–	–	10	1
16	2.25	0.6	100	–	–	–	–	1.5
17	2.25	0.6	90	10	–	–	–	1.5
18	2.25	0.6	90	–	10	–	–	1.5
19	2.25	0.6	90	–	–	10	–	1.5
20	2.25	0.6	90	–	–	–	10	1.5

ers. Each layer was tamped 25 times using a hard rubber mallet. The sample surfaces were finished using a metal spatula. After 24 hr, the specimens were demoulded and cured in water at 20°C. The suitable propagation of fibers in matrix is illustrated in Figure 5.

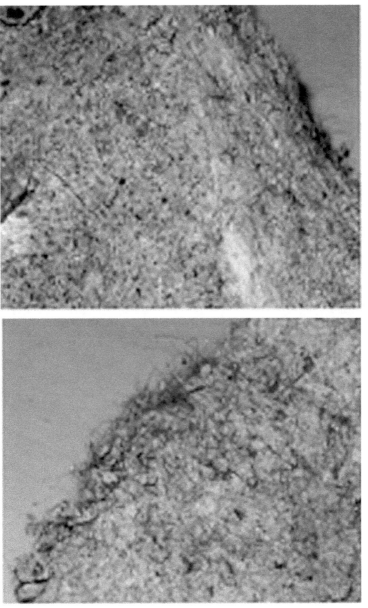

FIGURE 5   Porpagation of polypropylene fibers in mortar matrix (Left: 0.5% fiber and Right: 1% fiber).

The heating equipment was an electrically heated set. The specimens were positioned in heater and heated to desire temperature of 300 and 600°C at a rate of 10–12°C/min. After 3 hr, heater turned off. It was allowed to cool down before the specimens were removed to prevent thermal shock to the specimens. The rate of cooling was not controlled. The testes to determine the strength were made for all specimens at the age of 60 days. At least three specimens were tested for each variable.

16.4   DISCUSSIONS AND RESULTS

16.4.1   Density

The initial density of specimens containing polypropylene fibers was less than that of mixes without any fibers. Density of control mixes without any replacement of cement at 23,300 and 600°C are reported in Table 4. According to the results, density decrease of fiber reinforced specimens was close to that of plain ones. The weight of the melted fibers was negligible. The weight change of mortar was mainly due to the dehydration of cement paste.

TABLE 4   Density of control specimens.

Heated at	23°	300°	600°	PP fibers
Density (gr/cm³)	2.57	2.45	2.45	0%
Density (gr/cm³)	2.50	2.36	2.36	0.5%
Density (gr/cm³)	2.44	2.28	2.27	1%
Density (gr/cm³)	2.41	2.23	2.21	1.5%

16.4.2   Compressive Strength

In order to assess the effect of elevated temperatures on mortar mixes under investigation, measurements of mechanical properties of test specimens were made shortly before and after heating, when specimens were cooled down to room temperature. Compressive strength of reference specimen and heated ones at the age of 60 days are illustrated in Figure 6.

According to the results, by increasing the amount of polypropylene fibers in matrix the compressive strength of specimens reduced. Also, it is clear that the compressive strength of specimens were decreased by increasing the temperature to 300 and 600°C, respectively as supported by literatures.

The rate of strength reduction in fiber reinforced specimens is more than the plain samples and by rising the temperature it goes up.

The basic factor of strength reduction in plain specimens is related to matrix structural properties exposed to elevated temperature, but this factor for fiber reinforced specimens is related to properties of fibers. Fibers melt at temperature higher than 190°C and generate lots of holes in the matrix. These holes are the most important reasons of strength reduction for fiber reinforced specimens.

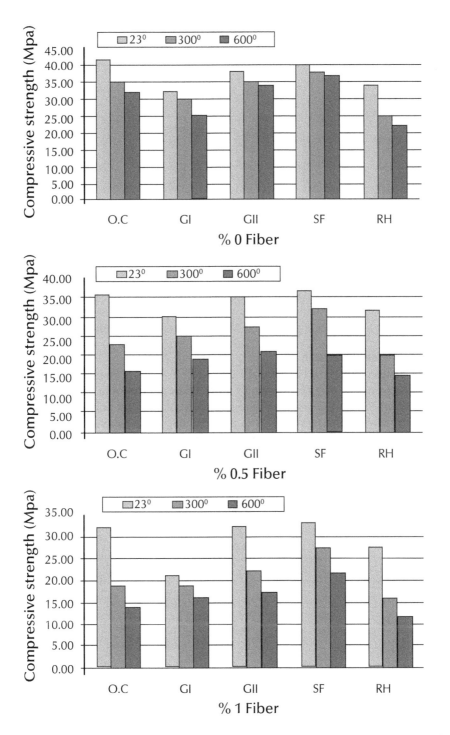

FIGURE 6   *(Continued)*

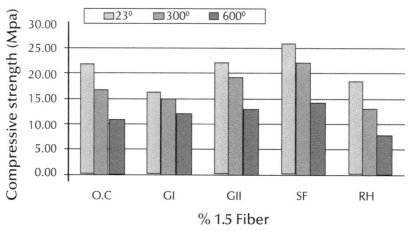

FIGURE 6   Compressive strength of samples at different temperatures.

Also, results indicate that silica fume and glass type II have an appropriate potential to apply as a partial replacement of cement due to their respective pozzolanic activity index values (according to ASTM C618 and C989, and Table 2).

16.4.3   Flexural Strength

The specimens were used for flexural testes were $50 \times 50 \times 200$ mm. The results of plain specimens and samples containing 1.5% fibers are shown in Table 5. The heat resistance of the flexural strength appeared to decrease when polypropylene fibers were incorporated into mortar. This is probably due to the additional porosity and small channels created in the matrix of mortar by the fibers melting like compressive strength. However, the effect of the pozzolans on flexural strength is not clear but it seems that silica fume and glass types II have better impact on strength in compare with control specimens than RH and glass type I.

TABLE 5   Flexural strength of control sampels with 0% and 1.5% fibers.

Batch	Flexural Strength (Mpa)			PP fibers
No	23°	300°	600°	(by volume)
1	4	3.1	2.8	0
2	3	2.4	2.1	0
3	3.7	3.2	2.7	0
4	3.9	3.4	2.7	0
5	3.3	2.8	2.6	0
16	2.2	1	0.7	1.5

TABLE 5 (*Continued*)

Batch	Flexural Strength (Mpa)			PP fibers
No	23°	300°	600°	(by volume)
17	1.6	0.8	0.6	1.5
18	2	1.1	0.8	1.5
19	2.4	1.3	0.8	1.5
20	1.8	1	0.6	1.5

16.5 CONCLUSION

This research proves the effects of elevated temperature on polypropylene fiber reinforced mortar as cement based composite and improving impacts of some pozzolanic materials like silica fume, RH, and especially finely ground glass on the strength of composite at high temperatures.

Based on the experimental results of this investigation the following conclusion can be drawn:

- Application of fibers in matrix causes the noticeable reduction in compressive and flexural strength.
- Mixture of cement based composite with GII and SF containing different percentage of fibers shown close mechanical properties to target specimens. So, results show the great usage possibility of ground glass and silica fume in composite as a partial replacement of cement.
- According to the results, there was a significant different between the compressive strength of specimens which include polypropylene fibers deal with plain specimens when expose to high temperature.
- In plain specimens by rising the temperature, size and rate of compressive strength reduction were not considerable. But results show that these difference rise up by increasing the fibers volume fraction in the matrix.
- The basic factor for strength reduction of fiber reinforced specimens is the melting of polypropylene fiber at temperature more than 190°C and generation of holes in the matrix.
- Also it is clear that increasing the temperature and fiber volume fracture in matrix, have a negative impact on density of specimens.

KEYWORDS

- **Construction technology**
- **Green house gas**
- **Mechanical biological treatment**
- **Municipal solid waste**
- **Polypropylene fibers**

REFERENCES

1. Wei, M. S. and Huang, K. H. Recycling and reuse of industrial waste in Taiwan. *Waste Management.*, **21**, 93–97 (2001).
2. Mildness, S., Young, J. F., and Darwin, D. *Concrete*. Prentice Hall, New Jersey (2003).
3. ACI Committee 232, Use of fly ash in concrete (ACI 232.2R-96), *ACI Manual of Concrete Practice, Part 1*, American Concrete Institute, Farmington Hills, (2001).
4. ACI Committee 232, Use of Raw or Processed Natural Pozzolans in Concrete (ACI 232.1R-00), American Concrete Institute, Farmington Hills, (2000).
5. ACI Committee 233, Ground, Granulated Blast Furnace Slag as a Cementitious Constituent in Concrete (ACI 233R-95). *ACI Manual of Concrete Practice, Part 1*. American Concrete Institute, Farmington Hills, (2001).
6. ACI Committee 234, Guide for Use of Silica Fume in Concrete (ACI 234R-96). *ACI Manual of Concrete Practice, Part 1*. American concrete Institute, Farmington Hills, (2001).
7. Harrison, W. H. Synthetic aggregate sources and resources, *Concrete*, **8**(11), 41–44 (1974).
8. Johnston, C. D. Waste glass as coarse aggregate for concrete. *Journal of Testing and Evaluation*, **2**(5), 344–350 (1974).
9. Meyer, C., Baxter, S., and Jin, W. Potential of waste glass for concrete masonry blocks. In *Materials for a New Millennium*. K. P. Chong (Ed.). Proceedings of ASCE Materials Engineering Conference, Washington, DC, pp. 666–673 (1996).
10. Polley, C., Cramer, S. M., and Cruz, R. V. Potential for using waste glass in Portland cement concrete. *Journal Materials in Civil Engineering*, **10**(4), 210–219 (1998).
11. Shao, Y., Lefort, T., Moras, S., and Rodriguez, D. Studies on concrete containing ground waste glass. *Cement and Concrete Research*, **30**(1), 91–100 (2000).
12. Shayan, A. and Xu, A. Value-added utilisation of waste glass in concrete. *Cement and Concrete Research*, **34**, 81–89 (2004).
13. Reindl, J. Report by recycling manager. *Dane County*. Department of Public Works, Madison (1998).
14. Park, S. B. and Lee, B. C. Studies on expansion properties in mortar containing waste glass and fibers. *Cement and Concrete Research*, **34**, 1145–1152 (2004).
15. Panchakarla, V. S. and Hall, M. W. Glascrete—disposing of non-recyclable glass. In *Materials for a New Millennium*, K. P. Chong (Ed.). Proceedings of ASCE Materials Engineering Conference, Washington, DC, pp. 509–518 (1996).
16. Meyer, C., Baxter, S., and Jin, W. Alkali-silica reaction in concrete with waste glass as aggregate. *Materials for a New Millennium*. In K. P. Chong (Ed.). Proceedings of ASCE Materials Engineering Conference, Washington, DC, pp. 1388–1394 (1996).
17. Panchakarla, V. S. and Hall, M. W. Glascrete—disposing of non-recyclable glass. In *Materials for a New Millennium*, K.P. Chong (Ed.). Proceedings of ASCE Materials Engineering Conference, Washington, DC, pp. 509–518. (1996).
18. Bentur, A. and Mindess, S. Fiber reinforced cementitious on durability of concrete. Barking Elsevier (1990).
19. Wang, Y. and Wu, H. C. Concrete reinforcement with recycled fibers. *Journal of Materials in Civil Engineering* (2000).
20. Flynn L. Auchey and Piyush K. Dutta Use of recycled high density polyethylene fibers as secondary reinforcement in concrete subje cted to severe environment. *Proceedings of the International Offshore and Polar Engineering Conference* (1996).
21. Wu, H. C., Lim, Y. M., and Li, V. C. Application of recycled tyre cord in concrete for shrinkage crack control. *Journal of Materials Science Letters*, **15**(20), 1828–1831 (October 15, 1996).
22. Naaman, A. E., Garcia, S., Korkmaz, M., and Li, V. C. Investigation of the use of carpet waste PP fibers in concrete. *Proceedings of the Materials Engineering Conference* **1**, 799–808 (1996).
23. Balaguru, P. N and Shah, S. P. *Fiber reinforced cement composites*. McGraw Hill Inc, New York, p. 367 (1992).

24. Kumar, S., Polk, M. B., and Wang, Y. Fundamental studies on the utilization of carpet waste, presented at the SMART (Secondary Materials & Recycled Textiles, An International Association) Mid-Year Conference, Atlanta, GA (July, 1994).
25. Terro, M. J. and Hamoush, S. A. Behavior of confined normal-weight concrete under elevated temperature conditions. *ACI Materials Journal*, **94**(2), 83–89 (1997).
26. Terro, M. J. and Sawan, J. Compressive strength of concrete made with silica fume under elevated temperature conditions. *Kuwait Journal of Science and Engineering*, **25**(1), 129–144 (1998).
27. Kalifa, P., Menneteau, F. D., and Quenard, D. Spalling and pore pressure in HPC at high temperatures. *Cement and Concrete Research*, **30**(12), 1915–1927 (2000).
28. Saad, M., Abo-El-Ein, S. A., Hanna, G. B., and Kotkata, M. F. Effect of temperature on physical and mechanical properties of concrete containing silica fume. *Cement and Concrete Research*, **26**(5), 669–675 (1996).
29. Xu, Y, Wong, Y. L., Poon, C. S., and Anson, M. Impact of high temperature on PFA concrete. *Cement and Concrete Research*, **31**(7), 1065–1073 (2001).
30. Kodur, V. K. R., Wang, T. C., and Cheng, F. P. Predicting the fire resistance behavior of high strength concrete columns. *Cement and Concrete Composites*, **26**(2), 141–153 (2004).
31. Poon, C. S., Azhar, S., Anson, M., and Wong, Y. L. Performance of metakaolin concrete at elevated temperatures. *Cement and Concrete Composites*, **25**(1), 83–89 (2003).
32. Savva, A., Manita, P., and Sideris, K. K. Influence of elevated temperatures on the mechanical properties of blended cement concretes prepared with limestone and siliceous aggregates. *Cement and Concrete Composites*, **27**(2), 239–248 (2005).

17 Application of Polymers in Construction Technology (PART II)

Effects of Jute/polypropylene fiber on reinforced soil

A. K. Haghi and G. E. Zaikov

CONTENTS

17.1 INTRODUCTION

The world wide cement production in 2007 was 2.77 billion tons [1]. Asia is the first producer (70%), followed by European Union countries (9.5%). Indeed, cement industry can be considered strategic in fact, from one side it produces an essential product in building and civil engineering for the construction of safe, reliable, long lasting buildings, and infrastructures. On the other side it is very important from the economic point of view (for example Indian cement industry is playing a very import role in the economic development of the country). However, cement industry environment wise is also responsible for a large use of not renewable raw materials (clays and calcium carbonate) and fossil fuels (e.g. clinker, the main cement constituent, is obtained at T = 1,500°C) resulting in heavy emissions of carbon dioxide (CO_2) in atmosphere. In fact, in 2006, the European cement industry used an energy equivalent of about 26 Mt of coal for the production of 266 Mt of cement [2-21] and it is estimated that 1 tone of CO_2 is emitted for each tone of cement produced. This induces cement industry to consider the possibility to introduce waste of different nature and origin in cement productive process. Two routes are currently taken into consideration: one involves the use of waste as alternative fuel the other considers waste as a new cement constituent. In this chapter, we will show how to convert the recycled wastes into wealth with particular application in construction industries.

17.2   EXPERIMENTAL

The soil stabilization is a process to improve certain properties of a soil to make it serve adequately an intended engineering purpose. The improving of the ground properties with various methods is a common case of geotechnical engineering. The aim of the soil stabilization is to decrease the consolidation and permeability capacity and to increase bearing and shear resistance capacity [22-32].

Natural fibers used at an appropriate length and amount can develop sufficient bond with the soil-cement to enhance the overall toughness of the composite. The slope of stress/strain curve (below yield point) denotes material's stiffness steeper the curve, the stiffer material, gradient is known as modulus of elasticity or Young's modulus.

The soil used in this experimental program is a common mason sand with a grain size distribution such that 100, 94, 56, 22, 9, and 5%, of the material passes the No. 10, 20, 40, 60, 120, and 200 sieves respectively. The coefficient of uniformity C_u and the coefficient of curvature, Cc, for this soil are 2.65 and 1.02, respectively.

Gradation and soil classification of samples is given in Figure 1 in terms of particle size distribution and soil classification.

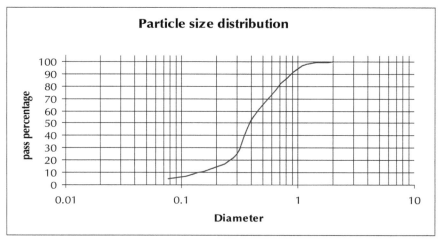

FIGURE 1   Particle size distribution.

The cement used is ordinary Portland cement. The physical and chemical properties of the cement are given in Table 1.

TABLE 1 Physical and chemical properties of the cement.

Physical properties	Cement
Fineness	3.12
Chemical composition	
Silica (SiO2)	20.44%
Alumina (Al2O3)	5.5%
Calcium oxide (CaO)	64.86%
Potash (K2O)	22.31%
Magnesia (MgO)	1.59%
Loss on ignition	1.51%
PH	12.06
3CaO. SiO2	66.48%
2CaO. SiO2	10.12%
4CaO. Al2O3. Fe2O3	9.43%
Free lime	1.65%
3CaO. Al2O3	8.06%

Jute yarn of 1100 tex yarn fineness was obtained from local firms. PP1 was supplied by LG as commercial grade HD 120 M. Jute yarn was dried in an oven at 100°C for 4 hr, then it was tested for moisture absorption by exposing the yarn to 50 and 95% RH atmosphere (in desiccators), at 23°C.

Brazilian tests were performed to obtain the tensile strength of specimens, based on ASTM C496 for indirect tensile test. The size and curing time of both tensile and compressive samples were similar. The load to failure was recorded and the tensile strength was computed as follows:

$$\sigma = \frac{2P}{\pi l d}$$

where, σ_t = Indirect tensile strength, P = Applied maximum load, l and d = length and diameter of the specimen, respectively.

The experiments were conducted with cement content varying jute/pp percentage (Figure 2 and Figure 3).

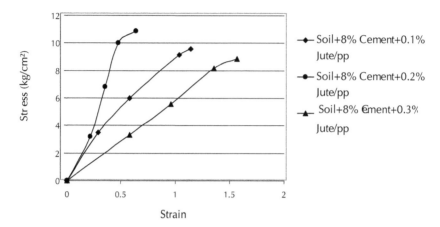

FIGURE 2   Effect of Jute/pp addition on stress-strain of raw soil-cement.

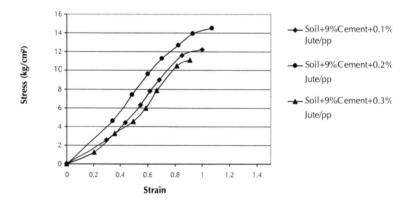

FIGURE 3   Effects of Jute/pp on stress-strain of raw soil-cement composite.

Figure 4 shows comparison of modulus of elasticity between different mixture conditions. Experimental results show that an increase in the percentage of cement content results in an increase in the modulus of elasticity. Also, adding fibers by one constant percent of cement results in an increase in the modulus of elasticity and toughness. Figure 4 shows that the maximum modulus of elasticity obtains from mixtures by 0.2% fibers content and it drops after the improvement of fibers from 0.2% (but it still is higher than similar rates in cases without fibers).

Indirect tensile strength tests were conducted on stabilized soil specimens containing 8, 9, and 10% of cement, non-reinforced and reinforced with 0.1, 0.2, and 0.3% of Jute/pp.

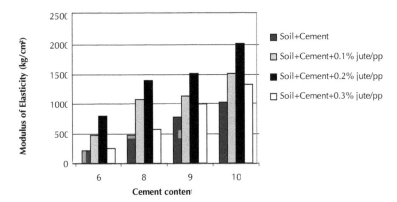

FIGURE 4 Comparison between modulus of elasticity of reinforced and non-reinforced specimens.

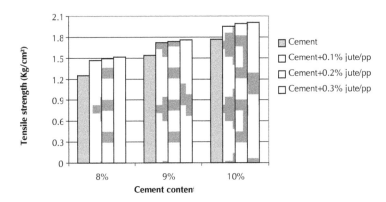

FIGURE 5 Effects of Jute/pp reinforcement on tensile strength of composite.

Two important parameters can be considered and discussed, when Jute/pp used as reinforcing fibers in cement stabilized sand samples. First of all, an increase in the tensile strength of specimens can be observed when specimens are reinforced with Jute/pp. Secondly, increase of cement content for a constant Jute/pp percentage increases the tensile strength of mixture.

The results of this experimental investigation in general indicate the importance of modulus of elasticity in evaluation of composite stiffness and interfacial bond between natural fiber and matrix.

17.3 BACKGROUND TO MICROMECHANICAL MODELS

The elastic properties of short fiber reinforced plastics can be experimentally determined or derived from a variety of mathematical models. The advantage of a comprehensive mathematical model is it reduces costly and time consuming experiments. Furthermore, a mathematical model may be used to find the best combination of constituent materials to satisfy material design considerations. Lastly, a physical (as opposed to empirical) model can yield insight into the fundamental mechanisms of reinforcement [33].

The purpose of the micro mechanical models is to predict the properties of a composite based on the properties of each constituent material [48]. Properties such as the elastic modulus Ec, Poisson's ratio (v) and the relative volume fractions (V) of both fiber and matrix are the fundamental quantities that are used to predict the properties of the composite. In some cases, fiber aspect ratio and fiber orientation are also included [33].

The variation of the amount of fibers in a natural fiber composite can be successfully chosen to correlate with the mechanical properties of composite. The amount of fibers is one of the most important characteristics of any composite material since their mechanical properties are strongly dependent on it [34].

The volume fraction of fibers is commonly used to estimate certain mechanical properties of the composite material. Besides, the type of fibers distribution (aligned or random) and their mechanical properties as well as properties of resin should be available. The volume fraction of a composite is obtained by the following formula:

$$Vf=\rho m Wf / (\rho m Wf + \rho f Wm)$$

Where Wf is the fiber weight fraction, Wm the matrix weight fraction, ρf the density of the fibers and ρm the density of the matrix [26, 48-49].

The mechanical properties of a composite material depend primarily on the strength and modulus of the fibers, the strength and the chemical stability of the matrix and the effectiveness of the bonding between matrix and fibers in transferring stress across the interface. Generally, the utilization of natural fibers as reinforcing materials in thermoplastics requires strong adhesion between the fiber and the matrix. Cellulose has a strong hydrophilic character due to three hydroxyl groups per monomeric unit, but biopolymers are generally hydrophobic [27]. A large number of research interests were dedicated to theoretical and numerical models with varying degrees of success.

It is necessary to perform comparison of the methods in order to determine the best approach.

The following models are applied to green composites:

1.   Rule of mixture (ROM) [35]
2.   Inverse rule of mixture (IROM) [36]
3.   Cox's model [37]
4.   Halpin-Tsai equation [38, 39]
5.   Our proposed model

17.3.1   Rule of Mixture (ROM)

The simplest available model that can be used to predict the elastic properties of a composite material is the rule of mixtures (ROM). To calculate the elastic modulus of the composite material in the one-direction E_1, it is assumed that both the matrix and fiber experience the same strain. This strain is a result of a uniform stress being applied over a uniform cross sectional area. The ROM equation for the apparent Young's modulus in the fiber direction is:

$$E_1 = E_f V_f + E_m V_m$$

Where E_f, E_m, V_f and V_m are the module and volume fractions of the fiber and matrix materials respectively. This model works extremely well for aligned continuous fiber composites where the basic assumption of equal strain in the two components is correct [33, 40-47].

The modified representation of the ROM which was adopted to estimate the modulus of elasticity of a composite material with long randomly distributed fibers is as follows [48, 49]:

$$E_1 = \eta E_f V_f + E_m V_m \quad \text{Or} \quad E_1 = \eta E_f V_f + E_m \left(1 - V_f\right)$$

It takes into account the weakening of the composite due to fibers orientation and fiber length factors through introduced additional coefficient, $\eta < 1$. Some attempts by other researchers have been done in order to estimate η. For example, in [50] it is suggested to apply $\eta = 0.2$ for a composite reinforced with randomly oriented natural fibers [34].

17.3.2 Inverse Rule of Mixture

The elastic modulus of the composite in the two-direction (E_2) is determined by assuming that the applied transverse stress is equal in both the fiber and the matrix (Reuss's assumption) [51].

As result, E_2 is determined by an inverse ROM equation that is given as:

$$E_2 = \frac{1}{\dfrac{V_f}{E_f} + \dfrac{V_m}{E_m}} \quad \text{Or} \quad E_2 = \frac{1}{\dfrac{V_f}{E_f} + \dfrac{1 - V_f}{E_m}}$$

17.3.3 Cox's Model (Modified Rule of Mixture)

The Cox shear lag theory adds to the ROM the shear lag analysis, which includes a fiber length and a stress concentration rate at the fiber's ends. The model is described by Equation (2), in which the coefficient η is written as follows:

$$\eta = 1 - \frac{\tanh\left(\dfrac{\beta l}{2}\right)}{\left(\dfrac{\beta l}{2}\right)}$$

where, l is the length of fibers and β is the coefficient of stress concentration rate at the ends of the fibers, which can be described by the following Equation (9):

$$\beta = \frac{l}{r} \sqrt{\frac{E_m}{E_f(1+v)\ln\sqrt{\frac{\pi}{4V_f}}}}$$

where, v is Poisson's ratio of fibers and r is the fiber radius.

17.3.4 Halpin-Tsai Equation

The semi-empirical equations developed by Halpin and Tsai are widely used for predicting the elastic properties of SFRT. The following form of the Halpin and Tsai equation is used to predict the tensile modulus of SFRT [33]:

$$E_1 = E_m \left(\frac{1+\xi\eta V_f}{1-\eta V_f} \right)$$

In Equation (6) the parameter η is given as:

$$\eta = \frac{(E_f/E_m)-1}{(E_f/E_m)+\xi}$$

Where ζ in Equations (6) and (7) is a shape fitting parameter to fit the Halpin-Tsai equation to the experimental data. The significance of the parameter ζ is that it takes into consideration the packing arrangement and the geometry of the reinforcing fibers [33, 48, 49]

A variety of empirical equations for ζ are available in the literature, and they depend on the shape of the particle and on the modulus that is being predicted [39]. If the tensile modulus in the principle fiber direction is desired, and the fibers are rectangular or circular in shape, then ζ is given by the following equation [39]:

$$\xi = 2\left(\frac{L}{T}\right) \qquad \text{Or} \qquad \xi = 2\left(\frac{L}{D}\right)$$

where, L refers to the length of a fiber in the one-direction and T or D is the thickness or diameter of the fiber in the three-direction. In Equation (8), as $L \to 0$, $\zeta \to 0$ and the Halpin-Tsai equation reduces to the IROM equation. In contrast when $L \to \infty$, $\zeta \to \infty$ and the Halpin-Tsai equation reduces to the ROM equation [33, 48].

17.3.5 Development of a new Model Suitable for Green Composites

A new theoretical model of the modulus estimation for natural fiber composite is necessary, since the existing models (at least those found in literature) cannot predict the Young modulus of composites with natural fibers in a reasonable error. A new model should be able to estimate reliably of the modulus of composite with different fiber volume fraction and with different elastic properties of fibers (with constant volume fraction) as well. On the other hand, the development of a new theoretical model from scratch is not reasonable, while combination of the existing models can be used as the basis for a new model.

During the benchmarking of the existing theoretical models it was found that Halpin model and IROM model give the modulus estimation quite close to the experimentally obtained data (Figure 6).

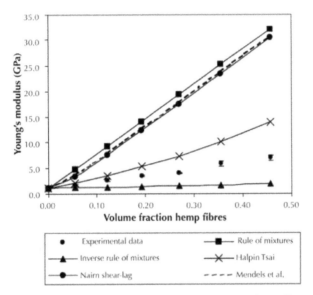

Determination of composite modulus containing hemp fibres.

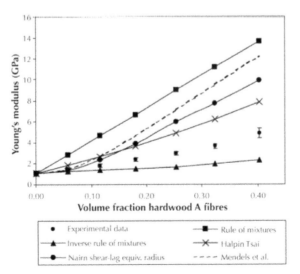

Determination of composite modulus containing hardwood A fibres.

FIGURE 6   Agreement of IROM and Halpin models with experimental data.

Therefore, the equations of these models can be used as a reference for the development of a new model. In composites with randomly distributed fibers there are fibers which are parallel, series and under an angle orientated to the chosen main direction. Unfortunately, ROM model does not take into account the influence of fibers which have orientation perpendicular to the chosen main direction, because in its equation the series part is absent. So the equation of this model has a linear behavior with respect to V_f fiber volume variable which does not give enough flexibility to adjust the model's behavior to interpolate the non-linear trend of experimentally obtained data. But because of great ability of ROM model in main direction, it can be used in conjunction with two other models say Halpin and IROM.

A try like this approach has been done in Ref [34], combining ROM and IROM models, with two new weight coefficients α and β:

$$E_c = \alpha \left[V_f E_F + \left(1 - V_f\right) E_m \right] + \beta \left(\frac{1}{V_f/E_f + \left(1 - V_f\right)/E_m} \right)$$

But a non scientific method has been used to find α and β coefficients.

The least squares methods with "r" and "s" criteria will be used to find the unique coefficients with the best convergence. The data for pp/Jute composite reinforced with randomly oriented fibers with different fiber volume fraction will be used, with $E_f = 41000 MPa$ and $E_m = 800 MPa$ [34].

Following models will be studied and compared:
- *Model 1*: IROM and modified ROM model (Figure 7).
- *Model 2*: modified ROM and Halpin (Figure 8).

TABLE 2   PP/Jute composite reinforced with randomly oriented fibers with different fiber volume fraction.

V_f	E_c, MPa
0	800
0.06	1300
0.12	1650
0.18	1800
0.24	2000
0.29	2100
0.34	2200
0.45	2250

Model 1:

$$E_1 = a\left(\eta E_f V_f + E_m \left(1 - V_f\right)\right) + b\left(\frac{1}{\dfrac{V_f}{E_f} + \dfrac{1 - V_f}{E_m}}\right)$$

In this model three weight coefficients, namely a, b, and η should be found in order to fit the data of Table 2 with best convergence.

With applying standard error and correlation coefficient criteria, a, b, and η constants are calculated as follows:

a =5.665136
b = – 4.5332004
η =0.05943155
Standard Error: 83.6682820
Correlation Coefficient: 0.9899705

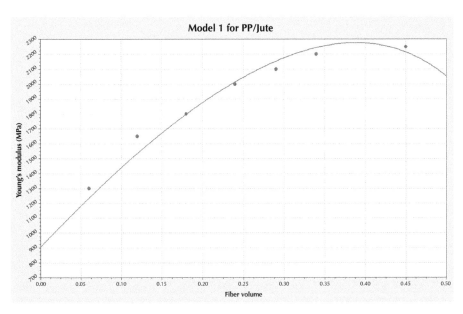

FIGURE 7   Data fit for IROM and modified ROM model.

Model 2:

$$E_1 = a \left(E_m \left(\frac{1 + \xi V_f \left(\dfrac{\left(E_f / E_m\right) - 1}{\left(E_f / E_m\right) + \xi} \right)}{1 - V_f \left(\dfrac{\left(E_f / E_m\right) - 1}{\left(E_f / E_m\right) + \xi} \right)} \right) \right) + b \left(\eta E_f V_f + E_m \left(1 - V_f\right) \right)$$

In this model four weight coefficients, namely a, b, ζ, and η should be found in order to fit the data of Table 2 with best convergence.

With applying standard error and correlation coefficient criteria, a, b, ζ, and η constants are calculated as follows:

$\eta = 0.65704045$

$\zeta = -4.9221785$

a = 0.82190093

b = 0.31977636

Standard Error: 97.3997226

Correlation Coefficient: 0.9891221

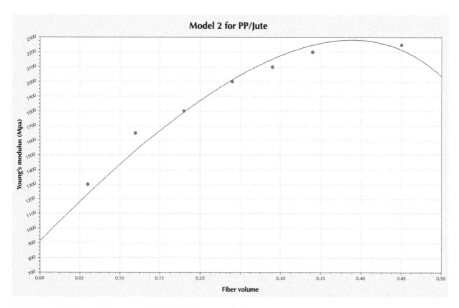

FIGURE 8   Data fit for modified ROM and Halpin.

17.4   MATHEMATICAL MODELS

When applying these approaches to choose a good model for fitting available data, some interesting properties of them has been considered and some pure mathematical models has been tested. Best series of models was sigmoidal family of models. Processes producing sigmoidal or "S-shaped" growth curves are common in a wide variety of applications such as biology, engineering, agriculture, and economics. These curves start at a fixed point and increase their growth rate monotonically to reach an inflection point. After this, the growth rate approaches a final value asymptotically. This family is actually a subset of the Growth Family, but is separated because of its distinctive behavior:

Gompertz model: $E = ae^{\left(-e^{\left(b-cV_f\right)}\right)}$

MMF model: $E = \dfrac{ab + cV_f^d}{b + V_f^d}$

Logistic model: $E = \dfrac{a}{1 + e^{\left(b - cV_f\right)}}$

Richards model: $E = \dfrac{a}{1 + e^{\left(b - cV_f\right)^{\frac{1}{d}}}}$

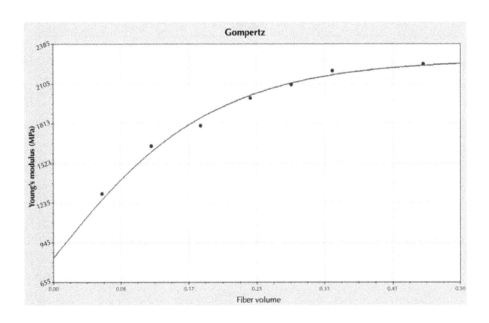

FIGURE 9   *(Continued)*

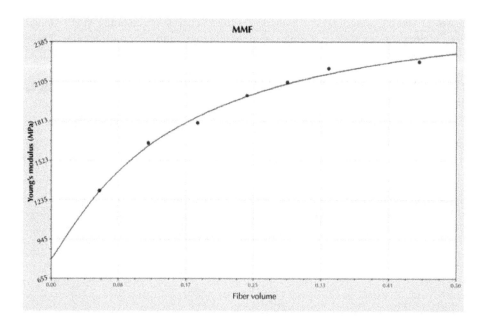

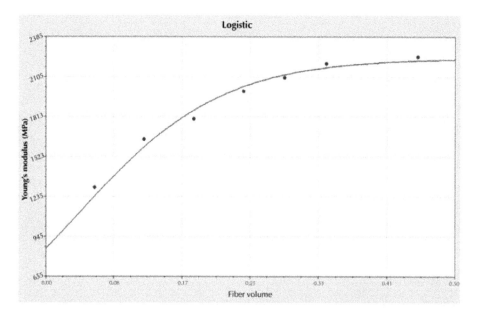

FIGURE 9   *(Continued)*

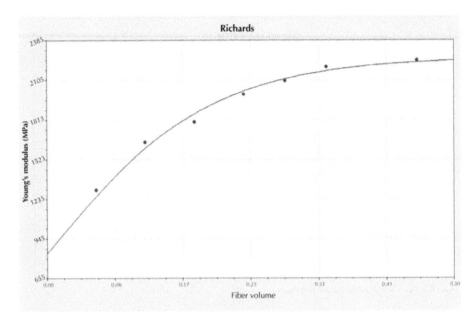

FIGURE 9 Data fit and comparison of models.

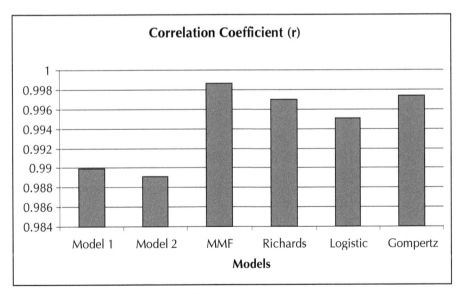

FIGURE 10 Models correlation coefficients.

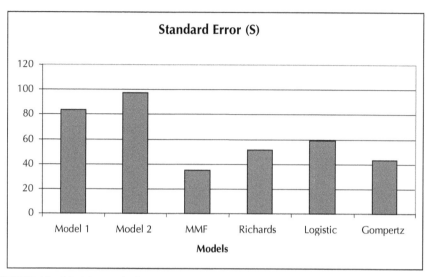

FIGURE 11   Models standard errors.

17.5   DISCUSSION AND RESULTS

The results of this study in general indicated that:

(1) The alternative base course composite has desirable strength and mechanical characteristics to be considered as a good quality stabilized pavement material.

(2) The higher the percentage of cement added, the higher the increment in strength and stiffness of treated soil.

(3) The compressive strength of the mixture increases with the increase of Jute/pp content until it reaches to its optimum value and then the additional fiber more than optimum value has decreasing effect on compressive strength.

(4) The maximum modulus of elasticity was obtained from mixtures by 0.2% Jute/pp.

(5) Reinforcing cement stabilized materials with Jute/pp improved the durability of soil cement mixtures.

(6) Reinforcing cement stabilized materials with Jute/pp improved indirect tensile strength of soil cement mixtures.

(7) Although our numerical investigation yields results that conform to the expected trends we explained, more research and modifications are necessary to be able to use it confidently. After the period spent studying this particular topic, we believe there is much potential for success in developing a more accurate model. It would be a time consuming task, but foundations are laid and the benefits are truly worthwhile.

17.6   CONCLUSION

A method for predicting the elastic modulus in green composites interfaces was developed. Through theoretical examinations a new model developed to estimate reliably of the modulus of elasticity in green composite interfaces with different fiber volume fraction and elastic properties. This approach allows a simple model for systems without resorting to complicated constitutive equations. The approach presented here, leads to theoretical predictions which can reasonably be explained from the physical point of view. Clearly, the final verification can only reached by systematic experimental investigations which, at present, are being carried out.

APPENDIX

Two criteria were adopted to evaluate the goodness of fit of each model, the Correlation Coefficient (r) and the Standard Error (S).

The standard error of the estimate is defined as follows:

$$S = \sqrt{\frac{\sum_{i=i}^{n_{points}} (E_{exp,i} - E_{pred,i})^2}{n_{points} - n_{param}}}$$

where, $E_{exp,i}$ is the measured value at point i, and $E_{pred,i}$ is the predicted value at that point, and n_{param} is the number of parameters in the particular model (so that the denominator is the number of degrees of freedom).

To explain the meaning of correlation coefficient, we must define some terms used as follow:

$$S_t = \sum_{i=1}^{n_{points}} (\bar{y} - E_{exp,i})^2$$

where, the average of the data points ($\bar{y}$) is simply given by:

$$\bar{y} = \frac{1}{n_{points}} \sum_{i=1}^{n_{points}} E_{exp,i}$$

The quantity S_t considers the spread around a constant line (the mean) as opposed to the spread around the regression model. This is the uncertainty of the dependent variable prior to regression. We also define the deviation from the fitting curve as

$$S_r = \sum_{i=1}^{n_{points}} (E_{exp,i} - E_{pred,i})^2$$

Note the similarity of this expression to the standard error of the estimate given: this quantity likewise measures the spread of the points around the fitting function. In view, the improvement (or error reduction) due to describing the data in terms of a regression model can be quantified by subtracting the two quantities. Because the magnitude of the quantity is dependent on the scale of the data, this difference is normalized to yield

$$r = \sqrt{\frac{S_t - S_r}{S_t}}$$

where, r is defined as the correlation coefficient. As the regression model better describes the data, the correlation coefficient will approach unity. For a perfect fit, the standard error of the estimate will approach $S = 0$ and the correlation coefficient will approach $r = 1$. [52]

The standard error and correlation coefficient values of all models are given in Figure 10 and Figure 11.

KEYWORDS

- **Construction technology**
- **Correlation coefficient**
- **Halpin-Tsai equation**
- **Micromechanical models**
- **Soil stabilization**

REFERENCES

1. Joseph, K., Thomas, S., and Pavithran, C. Effect of chemical treatment on the tensile properties of short sisal fiber-reinforced polyethylene composites. *Polymer*, **37**, 5139–5145 (1996).
2. Varma, I. K., Ananthakrishnan, S. R., and Krishnamoorthi, S. Comp of glass/modified jute fabric and unsaturated polyester. *Composites*, **20**, 383 (1989).
3. Geethamma, V. G., Thomas Mathew, K., Lakshminarayanan, R., and Thomas, S. Composite of short coir fibers and natural rubber: effect of chemical modification, loading and orientation of fiber. *Polymer*, **39**, 1483 (1998).
4. Sreekala, M. S., Kumaran, M. G., and Thomas, S. Oil palm fibers: morphology, chemical composition, surface modification and mechanical properties. *J Appl Poly Sci.*, **66**, 821–835 (1997).
5. Pothan, L. A., Oommen, Z., and Thomas, S. Dynamic mechanical analysis of banana fiber reinforced polyester composites. *Composite Science and Technology*, **63**, 283–293 (2003).
6. Aklonis, J. J. and MacKnight, W. J. *Introduction to polymer viscoelasticity*. Wiley, New York (1983).
7. Murayama, T. *Dynamic mechanical analysis of polymeric materials*. Elsevier, New York (1978).
8. Ferry, J. D. *Viscoelastic properties of polymers and comp, vol 2*, Wiley, New York (1980).
9. Datta, C., Basu, D., and Banerjee, A. *J Appl Poly Sci.*, **85**, 2800–2807 (2002).
10. Pothan, L. A., Thomas, S. and Groeninckx, G. The role of fiber/matrix interactions on the dynamic mechanical properties of chemically modified banana fiber/polyester composites. *Compos. Part A: Appl. Sci. Manuf.*, **37**(9), 1260–1269 (2005).
11. Pothen, L. A., Thomas, S., and Neelakandan, N. R. *J Reinforced Plast Compos*, **16**, 744 (1997).
12. Joseph, S., Sreekala, M. S., Oommen, Z., Koshy, P., and Thomas, S. A comparison of the mechanical properties of phenol formaldehyde composites reinforced with banana fibers and glass fibers. *Composite Science and Technology*, **62**, 1857–1868 (2002).
13. Sanadi, A. R., Prasad, S. V., and Rohatgi, P. K. *Journal of Scientific and Industrial Research*, **44**, 437 (1985).
14. Roe, P. J. and Ansell, M. P. Jute reinforced polyester composites. *J Mater Sci.*, **20**, 4015 (1985).
15. Sanadi, A. R., Prasad, S. V., and Rohatgi, P. K. *J. Mater Sci.*, **21**, 81 (1986).
16. Heijenrath, R. and Peijs, T. *Advanced Compos Lett.*, **5**(3), 81 (1996).
17. Marchovich, N., Reboredo, M., and Aranguren, M. *J Appl Poly Sci.*, **61**, 119 (1996).

18. Zarate, C. N., Aranguren, M. I., and Reboredo, M. M. *J Appl Poly Sci.*, **77**, 1832 (2000).
19. Sreekala, M. S., George, J., Kumaran, M. G., and Thomas, S. The mechanical performance of hybrid phenol-formaldehyde-based composites reinforced with glass and oil palm fibers. *Composite Science and Technology*, **62**, 339–353 (2002).
20. George, S., Neelakantan, N. R., Varghese, K. T., and Thomas, S. Dynamic mechanical properties of isotactic polypropylene/nitrile rubber blend: Effects of blend ratio, reactive compatibilization, and dynamic vulcanization. *J Poly Sci. Part B Polymer Physics*, **35**, 2309–2371 (1997).
21. Varghese, H., Bhagawan, S. S., Someswara, Rao., and Thomas, S. Morphology, mechanical and dynamic mechanical properties of blends of nitrile rubber and ethylene vinyl acetate copolymer. *Eur. Poly J*, **31**(10), 957–967 (1995).
22. George, J., Bhagawan, S. S., and Thomas, S. Thermogravimetric and dynamic mechanical thermal analysis of pineapple fiber reinforced polyethylene composites. *J Ther. Anal.*, **47**, 1121–1140 (1996).
23. Varghese, S., Kuriakose, B., and Thomas, S. Mechanicaland viscoelastic properties of short sisal fiber reinforced natural rubber composites: effects of interfacial adhesion, fiber loading and orientation. *J Adhes. Sci. Technol.*, **8**, 234 (1994).
24. Joseph, K., Pavithran, C., and Thomas, S. Dynamic mechanical properties of short sisal fiber reinforced polyethylene composites. *J Reinf. Plast. Comp.*, **12**, 139 (1993).
25. Mohanty, A. K., Wibowo, A., Misra, M., and Drzal, L. T. Effect of process engineering on the performance of natural fiber reinforced cellulose acetate biocomposite. *Cpmposite part A*, **350**, 363–370 (2004).
26. Loan. Doan, T. T., Gao, Sh. L., and Ma‥der, E. Jute/polypropylene composites I. Effect of matrix modification. *Composite Science and Technology*, **66**, 952–963 (2006).
27. Huda, M. S., Mohanty, A. K., Drzal, L. T., Schut, E., and Misra, M. "Green" composites from recycled cellulosend poly(lactic acid): Physico-mechanical and morphological properties evaluation. *Journal of Materials Science,* **40**, 4221–4229 (2005).
28. Loan. Doan, T. T. and Ma‥der, E. Performance of jute fiber reinforced polypropylene. In 7th international AVK-TV conference, Baden-Baden (2004).
29. Lucka, M., Bledzki, A. K., and Michalski, J. Influence of the hydrophobisation of Flax fibers on the water sensitivity, biological resistance and electrical properties of Flax–polypropylene composites. In *5th global wood and natural fiber composites symposium*, Kassel, Germany, p. 140 (April 27–28, 2004).
30. Gao, S. L., Ma‥der, E., and Plonka, R. Surface flaw sensitivity of glass fibers with carbon nanotube/polymer coating. In *International conference on composite materials (ICCM-15)*, Durban, South Africa, p. 962 (2005). Doan. T. T. Loan, *et. al.*, *Composites Science and Technology*, **66**, 952–963 (2006).
31. Cichocki, Jr, F. R. and Thomason, J. L. Thermoelastic anisotropy of a natural fiber. *Composite Science and Technology*, **66**, 669–678 (2002).
32. Haghi, A. K., Arabani, M., and Veis Karami, M. Applications of expanded polystyrene (EPS) beads and polyamide 66 in civil engineering, Part-2: Stabilization of clayey sand by lime/polyamide-66. *Composite Interfaces,* **13**(4–6), 451–459 (2006).
33. Facca, A. G., Kortschot, M. T., and Yan, N. Predicting the elastic modulus of natural fiber reinforced thermoplastics, Available online at www.elsevier.com/locate/compositesa.
34. Kavelin, K. G. Investigation of natural fiber composites heterogeneity with respect to automotive structures, ISBN-10: 90-9020036-3.
35. Sih, G. C., Carpinteri, A., and Surace, G. *Advanced Technology for Design and Fabrication of composite Meterials and Structures*. Kluwer Academic publishers (1995).
36. Jones, R. M. *Mechanics of composite material*. Hemisphere Publishing Corporation, United States, p. 86 (1975).
37. Cox, H. L. The elasticity and strength of paper and other fibrous materials. *Brit J Appl. Phys.*, **3**, 772–791 (1952).
38. Halpin, J. C. and Tai, S. W. *Effects of environmental factors on composite materials*, AFML-TR 7-423 (June, 1969).

39. Halpin, J. C. and Kardos, J. L. The Halpin–Tsai equations: A review. *Polym Eng Sci.*, **16**(5), 344–52 (1976).
40. Tsai, S. W. and Hahn, H. T. *Introduction to composite materials.* Technomic Publishing Co., Lancaster (PA) (1980).
41. Gibson, R. F. *Principles of composite material mechanics.* McGraw-Hill Inc, (1994).
42. Hyer, M. W. *Stress analysis of fiber reinforced composite materials,* McGraw-Hill Inc, (1998).
43. Daniel, I. M. and Ishai, O. *Engineering mechanics of composite materials.* University Press, Oxford (1994).
44. Herakovich, C. T. *Mechanics of fibrous composites.* John Wiley (1998).
45. Saha, A. K. Das, S., Bhatta, D., and Mitra, B. C. Study of jute fiber reinforced polyester composites by dynamic mechanical analysis. *J Appl Polymer Sci.*, **71**, 1505–1513 (1999).
46. Gassan, J. and Bledski, A. K. Possibilities for improving the mechanical properties of jute/epoxy composites by alkali treatment of fibers. *Compos Sci. Technol.*, **59**, 1303–1309 (1999).
47. Bergander, A. and Salmen, L. The transverse elastic modulus of the native wood fiber wall. *J Pulp Paper Sci.*, **26**, 234–238 (2000).
48. Bjurenstedt, A. and Lärneklint, F. *3D Biocomposites for Automotive Interior Parts.* MSc Thesis, Department of Applied Physics and Mechanical Engineering, Division of Polymer Engineering, Universitet ON, Luleå Tekniska (2004).
49. André, A. *Fibers for Strengthening of Timber Structures*, MSc Thesis, Department of Civil and Environmental Engineering, Division of Structural Engineering, Luleå Tekniska University of Technology, ON (2006).
50. Zárate, C. N., Aranguen, M. I., and Reboredo, M. M. Resol-vegetable fibers composites. *J Appl. Polym. Sci.*, **77**, 1832–1840 (2000).
51. Tucker, C. L. and Liang, E. Stiffness predictions for unidirectional short fiber composites: review and evaluation. *Comp Sci Technol.*, **59**, 655–671 (1999).
52. Daghbandan, A., Hajiloo, A., Farjad, A., and Haghi, A. K. Analytical *Determination of the Drying Characteristics.* 4th Asia Pacific Drying Conference Proceeding, pp. 802–815 (ADC 2005).
53. Nairn, J. A. On the use of shear-lag methods for analysis of stress transfer in unidirectional composites. *Mech. Mater*, **26**, 63–80 (1997).
54. Mendels, D. A., Leterrier, Y., and Manson, J. A. E. Stress transfer model for single fiber and platelet composites. *J Comp. Mater*, **33**(16), 1525–1543 (1999).

18 Application of Polymers in Construction Technology (PART III)

Effects of recycled Polyamide fibers and Expanded Polystyrene on cement composites

A. K. Haghi and G. E. Zaikov

CONTENTS

18.1 INTRODUCTION

Using waste as new cement constituent, it must be ascertained the absence of substances that can negatively interfere with cement reaction. European standard EN 197-1, providing a complete classification of cements valid in Europe, confirms that a constituent of cement can also be a waste. In fact, except the ordinary Portland cement (OPC) constituted by 95% of clinker (CEM I), blended cements contain, besides clinker, other constituents coming from waste of different productive processes. Blast furnace slag, silica fume and fly ash respectively derive from iron ore processing forecast iron and steel production, from silicon and ferrosilicon alloys production and from coal combustion for electric energy production. It has been proved that the addition of waste leads to several advantages for the relevant cements: (i) the lower content of clinker compare to OPC allows to classify these binders as low heat cements: (ii) the development of cement mechanical strength, although it is slower at the early curing age, noteworthy increases with time, even after 28 days: (iii) durability behavior improves. Moreover, blended cements always need less clinker for their production thus involving a minor use of natural raw materials and fuels, less quarries exploitation and lower CO_2 emission. When the new cement constituents are waste, benefits as the safe-guard of disposal sites and saving of natural raw materials must be added at the quoted list. In this chapter, we will show how to convert the recycled wastes into wealth with particular application in construction industries.

18.2   EXPERIMENTAL WORK

18.2.1   Materials

The OPC was used. The main chemical composition of (OPC) used in this study are listed in Table 1.

TABLE 1   Chemical and physical compositions of cement used in this study.

SiO_2	20.40%
Al_2O_3	6.12%
Fe_2O_3	3.051%
CaO	63.16%
MgO	2.32%
SO_3	2.40%
Specific gravity	3.13

Two types of commercially available spherical expanded polystyrene (EPS) beads were used (Table 2). As indicated in Table 2, the grading of EPS shows that Type A has mostly 6.0 mm size beads and Type B has mostly 3.0 mm size beads.

TABLE 2   EPS beads characteristics.

Type	Size mm	Bulk density Kg/m³	Specific gravity
A	6	16	0.014
B	3	20	0.029

The physical properties of PA 66 used in this study are summarized in Table 3.

TABLE 3   Some physical properties of PA 66.

Specific gravity	1.16
Tensile strength	965 MPa
Elastic modulus	

18.2.2   Admixture

A naphthalene based superplasticizer (that complies with ASTM C 494) was used to produce the mixes a flowable or highly workable nature, to suit the hand compaction adopted. This admixture is well known in the market and has a good history in international concrete works worldwide.

18.2.3   Specimen Preparation

A number of standard test specimens of different sizes were selected for investigating the various parameters. For studying the compressive strengths at 3, 7, and 28 days and also for investigating the absorption tests cubes of 100 mm size were used. Meanwhile, the split tensile strength test was conducted on 100 mm diameter × 200 mm cylinders at 28 days.

Lightweight EPS concrete were made with PA and without PA reinforcement. Initially, the EPS beads were wetted with 25% of the mixing water and superplasticizer before adding the remaining materials. It is expected that by addition of EPS in place of normal aggregate the weight of concrete can be reduced significantly and the interaction of Polyamide 66 fibers with EPS concrete can improve the strength. In this study, the estimated component additions were measured by volume in order to simplify mixing process. The non-absorbent, hydrophobic and closed cellular aggregates (expanded polystyrene beads) were mixed in a planetary mixer in compliance with the recommendation of ASTM C 305. To obtain a very uniform and flowing mixture, the mixing operation was continued on regular biases. The molds are then filled with fresh EPS concrete and then with PA reinforced EPS concrete. Afterward, the mixture is firmly compacted by hand.

Based on ASTM C 143-98 we measured the slump values of the fresh concrete. Afterward, the compressive strength test was carried out in a testing machine with capacity of 2,000 kN. The loading rate used during this test was 2.5 kN/s. According to ASTM C 496-89 we conducted the split tensile strength test on cylinders at 28 days. The absorption test was then carried out as per ASTM C 642-82.

18.3   EXPERIMENTAL DISCUSSION AND RESULTS

It appeared that the workability of the concrete in terms of the slump measurements were about 45–80 mm in non-reinforced cases and around 40–75 for polyamide EPS concretes. In addition, it should be noted that the mixes having the higher percentage silica fume can present higher slump values. During the experimental study, all the concretes specimens shown an interesting flexibility and provided a very easy condition to work with. Meanwhile, the specimens could be compacted using just hand compaction and they could be easily finished. In order to be able to solve the problems associated with the hydrophobic nature of the EPS beads, the silica fume, and the superplasticizer were added. This will help to improve the cohesiveness of the mix significantly.

Figure 1 displays the development of compressive strength with the age for reinforced EPS concrete. This illustration clearly states that the compressive strength of reinforced EPS concretes in almost all mixes show a continuous increase with age. It is seen that the rate of strength development was greater initially and decreased as the age increased. Comparison of strengths at 7 days revealed that non-reinforced concretes developed almost 40–50% of its 28 day strength. While it is seen that reinforced concretes developed almost 20–40% of the corresponding 28 day strength.

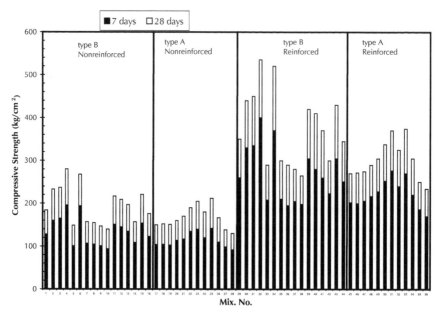

FIGURE 1   Effects of age on compressive strength.

Figure 2 and Figure 3 illustrate the compressive strengths of EPS concretes with different plastic densities for reinforced and non-reinforced concrete and the volume content of EPS.

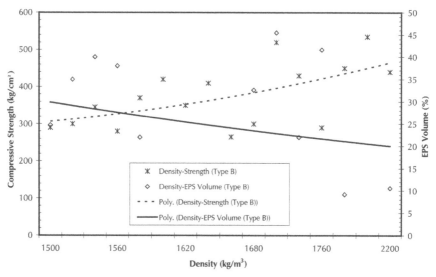

FIGURE 2   Variation of composite density with EPS volume and compressive strength (Type B).

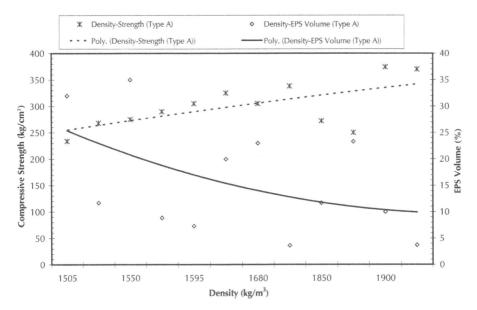

FIGURE 3   Variation of composite density with EPS volume and Compressive strength (Type A).

It is appeared that the strength of non-reinforced and reinforced EPS concrete to increase linearly with a decrease in the EPS volume. Furthermore, it is noted that the density of EPS concretes decreased significantly with an increase in the EPS volume. It is clearly observed that the polyamide EPS concrete have high strength in compression, comparing to that non-reinforced concretes.

In some earlier studies reported on EPS concrete, the densities varied from 1,500 to 2,000 kg/m³. At 28 days, it is reported a compressive strength of 100 to 210 kg/cm² for these concretes. In the present study, we obtained the strength of (235 to 375) and (265 to 540) with use of polyamide 66 yarns.

The plot represented in Figure 4 show an inverse relationship between the concrete strength and the bead size. That means the strength in EPS concrete increased with a decrease in the EPS bead size for the same mix proportions.

From Figure 5 it can be clearly appreciated that in the mixes content fine silica fume the compressive strength can increase significantly (i.e. mixes no. 4, 6,…). This represents the effects of fine silica fume and polyamide 66 on the compressive strength of EPS concrete. This is clearly visible that fine silica fume can improve the dispersion of EPS in the cement paste and interfacial bonding between EPS and cement paste. Moreover, the compressive strength of polyamide EPS concrete is almost 1.5–2 times non-reinforced EPS concretes.

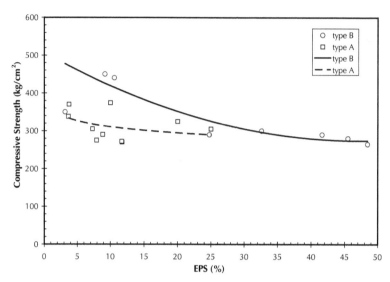

FIGURE 4   Effects of EPS bead size on compressive strength.

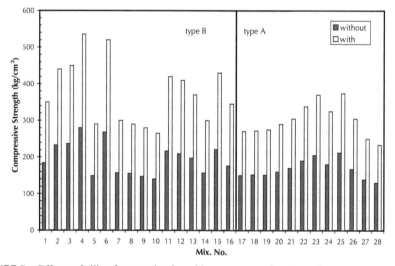

FIGURE 5   Effects of silica fume and polyamide on compressive strength.

The failure modes of different specimens are presented in Figure 6. Figure 6(a) shows the EPS concrete specimen before loading. The failure mode observed in Figure 6(b) is rather gradual (more compressible), and the specimen is capable of retaining the load after failure, without full disintegration. Figure 6(c) illustrates failure mode in polymer concretes containing EPS with polyamide yarns. This is attributed to the fact that the failure mode of this specimen is near to a typical brittle failure.

a)

b)

FIGURE 6 *(Continued)*

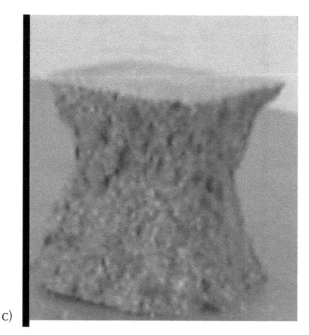

c)

FIGURE 6 The EPS concrete samples with and without polyamide 66 (a) Compressive samples before failure, (b) Observed failure mode in polymer concretes containing EPS without polyamide yarns and (c) Observed failure mode in polymer concretes containing EPS with polyamide yarns.

Figure 7 illustrates the effects of volume content of EPS on the tensile strength of polyamide EPS concretes. This attribute to the fact that there is an inverse relationship between the concrete tensile strength and the EPS volume. This means that the tensile strength of EPS concrete appeared to increase with a decrease in the EPS volume.

From Figure 8, the tensile strength of polyamide EPS concrete (Type B) is 40 kg/cm² whereas, the maximum tensile strength for polyamide EPS concrete (Type A) is only 31 kg/cm². But in general, we can expect an increase of 70% in the tensile strength for the lower bead sizes.

The effect of polyamide 66 yarns on the tensile strength of EPS concrete is presented in Figure 8. The tensile strength of Polyamide-EPS concrete demonstrated a higher value with respect to EPS concrete. This incremental ratio of strengths is about 60 to 100%. Furthermore it is observed that the Polyamide EPS concrete containing smaller beads have higher tensile strength.

Efforts were made to study the variation of tensile strength with the compressive strength (Figure 9). It appears that the tensile strength increased with an increase in compressive strength. The rate of strength increasing in polyamide EPS concrete containing small beads (Type B) is higher than other.

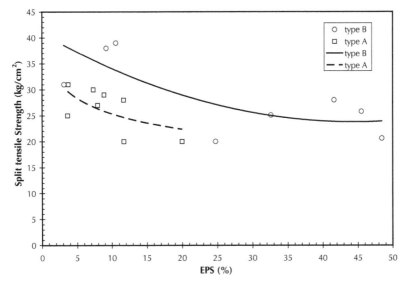

FIGURE 7   Variation of tensile strength with EPS volume.

The splitting failure mode of the concrete specimens containing non-reinforced EPS aggregates also did not exhibit the typical brittle failure principally expected in conventional concrete as in compressive strength. Consequently, the tensile failure mode observed in the polymer concrete containing polyamide 66 yarns was rather a typical brittle failure which is normally resembles to conventional concrete (Figure 10).

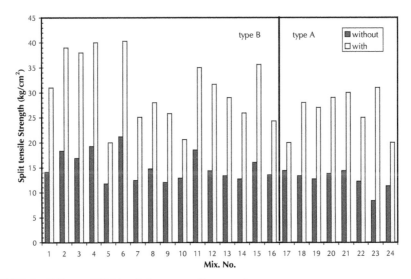

FIGURE 8   Effects of EPS bead size on tensile strength.

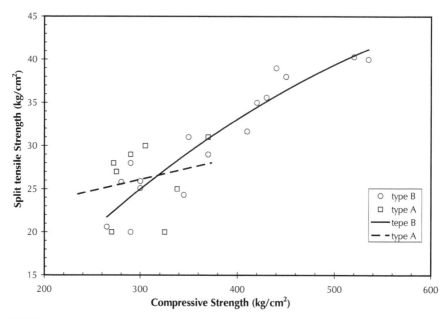

FIGURE 9   Variation of tensile strength with compressive strength.

FIGURE 10   The failure mode in split tensile test.

The most significant factor affecting the shrinkage of concrete is the degree of restraint by the aggregate (i.e. its elastic properties and the volumetric proportion of the paste in the mix). It should be noted that the EPS beads principally offer little hindrance to the shrinkage of the paste. Therefore, it is expected that as the volumetric proportion of the EPS is increased, the shrinkage would increase as well. This is clearly represented in Figure 11. In this study, the drying shrinkage of normal concrete at 90 days was 610 microstrain. Inversely, when the volume content of EPS in the concrete specimen was 20%, the drying shrinkage at 90 days raised to 1,000 microstrain, which represents a disadvantage for the application of EPS.

In a view of, it can be seen that polyamide yarns improved the drying shrinkage of EPS concrete significantly. Even for the case of EPS concrete which contained 20% of its volume by EPS, the drying shrinkage at 90 days was 670 microstrain, which seems to be rather close to normal concrete range.

Figure 12 represent the final absorption for polyamide reinforced concretes with different EPS volume. The initial and final absorption for these concretes using Type A and B EPS is illustrated in Figure 13. From these illustrations, it can be observed that the initial surface absorption of all the EPS concretes are as low as 2.5%, which is specified as "good" concrete by CEB [9]. Nevertheless, the final absorption for these concretes also followed a similar trend. Moreover, the total absorption values observed ranged from 2 to 4.5% for specimens. In general, from the results it can be seen that, lower strength specimens were showing higher absorption compared to the higher strength specimens. From Figure 3 and Figure 10, it is revealed that the mixes of 6, 7, have higher strength while their absorption is lower than others.

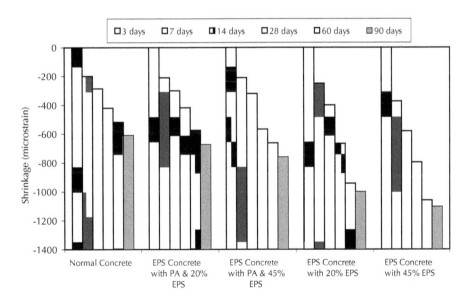

FIGURE 11 Relationship between drying shrinkage strain of EPS concrete (with polyamide and without polyamide) and reference concrete with age.

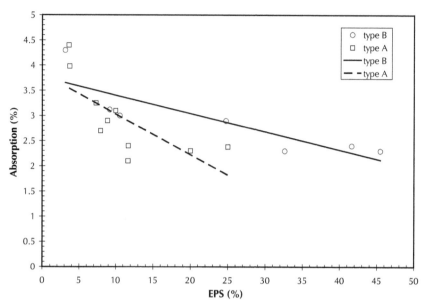

FIGURE 12   Absorption characteristics of EPS concretes.

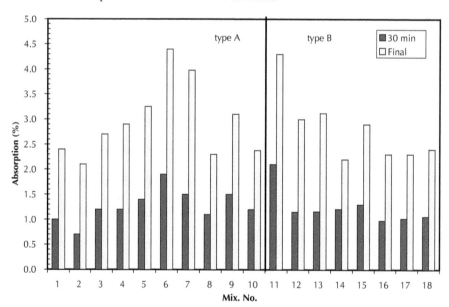

FIGURE 13   Comparison of initial and final absorption.

In this research the maximum range of absorption value obtained was 4.5%. This could be attributed to the fact that the application of polyamide 66 yarns had a significant influence on the absorption value.

18.4 CONCLUSION

Upon cracking, fibers are able to bridge the initial crack and hold the crack together and sustain the load until the fibers either pull-out from the matrix (in early age) or fracture that say flexural toughness. According to Figure 14 it can be observed that after crack initiation, fibers can still carry load and absorb the energy.

Figure 15 represents typical load-deflection responses of plain matrix and mixtures contained 0.5, 1, and 1.5% fibers.

FIGURE 14 *(Continued)*

FIGURE 14   Cracking of flexural specimens and bridging effect of fiber on cracks.

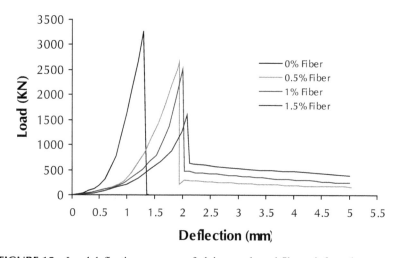

FIGURE 15   Load deflection response of plain sample and fiber reinforced.

For 0% fibers used, the behavior was in a brittle manner. When the strain energy was high enough to cause the crack to self-propagate, fracture occurred almost while the peak load was reached (this is due to the tremendous amount of energy being re-leased). According to Figure 15, the fiber bridging effect helped to control the rate of energy release significantly. Thus, fibers still can carry load even after the peak. With the effect of fibers bridging across the crack surface, fibers were able to maintain the load carrying ability even after the samples had been cracked. These are accordance to ASTM C1018, in which toughness or energy absorption defined as the area under the load deflection curve from crack point to 1/150 of span.

Our laboratory results indicate that, by increasing the amount of fiber in matrix flexural toughness rise up. The use of recycled materials in concrete has been on the increase throughout the world due to conserving resources. The results of this study have shown that use of this locally available low cost PA waste may provide an economical and effective alternative to concrete aggregates. The chief innovation of this work is the significant ability of polyamide 66 fibers to improve the behavior of lightweight EPS concrete. It is concluded that: The interaction of Polyamide 66 fibers with EPS concrete did significantly influence the strength. For a 28 days lightweight Polyamide-EPS concrete, we demonstrated strength of around 500 kg/cm^2 while the densities varied from 1,500 to 2,000 kg/m^3.

However, the tensile strength of Polyamide-EPS concrete demonstrated a value of about 40 kg/cm^2. This incremental ratio of strengths represents an improvement of about 60–100%. Meanwhile, final absorption of Polyamide-EPS concrete demonstrated a value of 4.5%. This value for EPS concrete was about 8.2%. It was noted that the drying shrinkage can be considered as the weakness of EPS concretes. This can be resolved by addition of Polyamide 66 fibers in EPS concrete.

In this study, Polyamide 66 fibers improved the drying shrinkage of EPS concrete significantly. Addition of Polyamide 66 fibers to those EPS concretes containing 20% of EPS demonstrated a drying shrinkage equal to normal concrete. But the presence of polyamide 66 yarns in EPS concretes containing 45% of EPS reduced the drying shrinkage by 30%.

KEYWORDS

- **Construction technology**
- **Ordinary Portland cement**
- **Polyamide yarns**
- **Reinforced EPS concrete**
- **Super plasticizer**

REFERENCES

1. Muthukumar, M. and Mohsen, D. Studies on polymer concrete based on optimized aggregate mix proportion. *European Polymer Journal*, **40**, 2167–2177 (2004).
2. Gorninski, J. P., Dal Mobin, D. C., and Kazmierczak, C. S. Study of the modulus of elasticity of polymer concrete compounds and comparative assessment of polymer concrete and Portland cement concrete. *Cement and Concrete Research*, **34**, 2091–2095 (2004).
3. ACI Committee 213 R-87. *Guide for Structural Lightweight Aggregate Concrete*, ACI Manual of Concrete Practice, Part 1, American Concrete Institute, Farmington Hills (1987).
4. Short, A. and Kinniburgh, W. *Lightweight Concrete*. (3rd ed.), Applied Science Publishers, London (1978).
5. Sussman, V. Lightweight plastic aggregate concrete. *ACI J*. pp. 321–323 (July, 1975).

6. Bagon, C. and Frondistou-Yannas, S. Marine floating concrete made with polystyrene expanded beads. *Mag. Concr. Res.,* **28**, 225–229 (1976).
7. Babu, K. G. and Babu, D. S. Behavior of lightweight expanded polystyrene concrete containing silica fume. *Cement and Concrete Research*, **33**(5), 755–762 (May, 2003).
8. Al-Manaseer, A. A. and Dalal, T. R. Concrete containing plastic aggregates. *Concr. Int.,* 47–52 (August, 1997).
9. CEB-FIP. Diagnosis and assessment of concrete structures—State of the art report, CEB Bulletin (1989).

19 Application of Polymers in Construction Technology (PART IV)

Effects of EPS size on Engineering Properties of Concrete

A. K. Haghi and G. E. Zaikov

CONTENTS

19.1 INTRODUCTION

Economic development is always accompanied by higher consumption of goods, services and attendant increased generation of solid wastes that need to be disposed somehow. The waste generation today is higher than the economic growth and different waste management methods aim to reduce the significant environmental and economic impact of this fact. In the generally accepted waste hierarchy, the first priority is for waste reduction, followed by recycling and also composting of clean biodegradable organic wastes (food and yard wastes). The EU promotes recycling over other waste treatment methods for recovering materials. In this way, physical resources are protected since paper, metals, glass, and plastics that are recovered from the waste stream demand less resources and energy than the use of "virgin" materials. Landfilling is the most common method for waste management in many EU Member States and in some cases this dependency exceeds 80%. The EU Landfill Directive of 1999 obliges Member States to progressively reduce the amount of organic waste going to landfill to 35% of the 1995 levels within 15 years aims to reduce such a loss of resources. This clear policy direction has put emphasis on waste management systems that increase and optimise the recovery of resources from waste whether as materials or as energy.

The production of municipal solid waste (MSW) is as old as mankind itself. Every civilization, from the oldest to the most modern economic society, has had to deal with MSW. Of course, the style of handling is different comparing the earliest civilizations to the 21st century. Not only is the amount of wastes greater, but the composition of wastes is completely different. The industrial revolution between 1750 and 1850 led many people to move from rural areas to cities. The concentration of the inhabitants of towns and cities caused an increase in waste amount. These wastes generated contained a range of materials such a glass, food residue and human waste. An additional problem was the attraction of MSW for flies, rats and other vermin which can cause disease transfer. These conditions were very dangerous, and caused an actual threat to human health and the environment.

Today, the composition and amount of MSW is extremely variable as a consequence of seasonal, lifestyle, demographic, geographic, and legislative factors. This variability makes defining and measuring the composition of waste more difficult and at the same time more essential. The MSW, also called urban solid waste, is a waste type that includes predominantly household waste (domestic waste), with sometimes the addition of commercial wastes, collected by a municipality within a given area. They are in either solid or semisolid form and generally exclude industrial hazardous wastes. The term residual waste relates to waste left over from household sources containing materials that have not been separated out or sent for reprocessing.

Waste management is the collection of transport, processing, recycling or disposal, and monitoring of waste materials. The term usually relates to materials produced by human activities, and is generally undertaken to reduce their effect on health, the environment or aesthetics. Waste management is also carried out to recover resources. Waste management can involve solid, liquid or gaseous substances, with different methods and fields of expertise for each.

Effective waste management through MSW composition studies is important for numerous reasons, including the need to estimate the potential of material recovery, to identify sources of component generation, to facilitate the design of processing equipment, to estimate physical, chemical, and thermal properties of waste, and to maintain compliance with national laws and European directives.

Moreover, the demand for lightweight concrete in many applications of modern construction is increasing, owing to the advantage that lower density results in a significant benefit in terms of load-bearing elements of smaller cross sections and a corresponding reduction in the size of the foundation [3].

Lightweight aggregate concrete, popular through the ages, was reported to have a comparable or some times better durability even in severe exposure conditions. Lightweight aggregates are broadly classified in to two types—natural (pumice, diatomite, volcanic cinders, etc.) and artificial (perlite, expanded shale, clay, slate, sintered PFA, etc.). Expanded polystyrene (EPS) beads are a type of artificial ultra-lightweight, nonabsorbent aggregate [4, 5]. It can be used to produce low-density concretes required for building applications like cladding panels, curtain walls, composite flooring systems, and load-bearing concrete blocks [6, 7].

The EPS concrete is a lightweight concrete with good energy absorbing characteristics, consisting a discrete air voids in a polymer matrix. However, polystyrene beads are extremely light, with a density of only $12\text{--}20 \text{ kg/m}^3$, which can easily cause segregation in mixing. Hence, some chemical treatment of surface on this hydrophobic material is needed. Other investigators also reported that EPS tends to float and can result in a poor mix distribution and segregation, necessitating the use of admixtures [8, 9].

19.2 SCOPE

Four key points of this investigation are:
(1) To convert carpet waste into useful product.
(2) To consume carpet wastes which would otherwise gone to landfill?
(3) Protection of Environment from being heavily contaminated.
(4) Using recycled Polyamide (PA 66) fibers could be a cost-effective way to improve the performance for lightweight EPS concrete.

19.3 EXPERIMENTAL

19.3.1 Materials

Ordinary Portland cement (OPC) was used. The main chemical composition of (OPC) used in this study are listed in Table 1.

TABLE 1   Chemical and physical compositions of cement.

SiO_2	**20.40%**
Al_2O_3	6.12%
Fe_2O_3	3.051%
CaO	63.16%
MgO	2.32%
SO_3	2.40%
Specific gravity	3.13

Two types of commercially available spherical EPS beads were used (Table 2). As indicated in Table 2, the grading of EPS shows that Type B has mostly 3.0 mm size beads and Type C has mostly 6.0 mm size beads

TABLE 2   EPS beads characteristics.

Type	Size mm	Bulk density kg/m³	Specific gravity
B	3	20	0.029
C	6	16	0.014

The physical properties of PA 66 used in this study are summarized in Table 3.

TABLE 3   Some physical properties of PA 66.

Specific gravity	1.16
Tensile strength	965 MPa
Elastic modulus	5.17 GPa
Ultimate elongation	20%

19.3.2   Admixture

A naphthalene based super plasticizer (that complies with ASTM C 494) was used to produce the mixes a flow able or highly workable nature, to suit the hand compaction adopted. This admixture is well known in the market and has a good history in international concrete works worldwide.

19.3.3   Specimen Preparation

A number of standard test specimens of different sizes were selected for investigating the various parameters. For studying the compressive strengths at 3, 7, and 28 days and also for investigating the absorption tests cubes of 100 mm size were used. Meanwhile, the split tensile strength test was conducted on 100 mm diameter × 200 mm cylinders at 28 days.

Lightweight EPS concrete were made with PA and without PA reinforcement. Initially, the EPS beads were wetted with 25% of the mixing water and super plasticizer before adding the remaining materials. It is expected that by addition of EPS in place of normal aggregate the weight of concrete can be reduced significantly and the interaction of Polyamide 66 fibers with EPS concrete can improve the strength. In this study, the estimated component additions were measured by volume in order to simplify mixing process. The non-absorbent, hydrophobic and closed cellular aggregates (expanded polystyrene beads) were mixed in a planetary mixer in compliance with the recommendation of ASTM C 305. To obtain a very uniform and flowing mixture, the mixing operation was continued on regular biases. The molds are then filled with fresh EPS concrete and then with PA reinforced EPS concrete. Afterward, the mixture is firmly compacted by hand.

Based on ASTM C 143-98 we measured the slump values of the fresh concrete. Afterward, the compressive strength test was carried out in a testing machine with capacity of 2,000 kN. The loading rate used during this test was 2.5 kN/s. According to ASTM C 496-89 we conducted the split tensile strength test on cylinders at 28 days. The absorption test was then carried out as per ASTM C 642-82.

19.4 EXPERIMENTAL, DISCUSSION, AND RESULTS

The results of this investigation are explained in the following subsections.

19.4.1 Fresh Concrete

It appeared that the workability of the concrete in terms of the slump measurements were about 45–80 mm in non-reinforced cases and around 40–75 for polyamide EPS concretes. In addition, it should be noted that the mixes having the higher percentage silica fume can present higher slump values. During the experimental study, all the concretes specimens shown an interesting flexibility and provided a very easy condition to work with. Meanwhile, the specimens could be compacted using just hand compaction and they could be easily finished. In order to be able to solve the problems associated with the hydrophobic nature of the EPS beads, the silica fume and the super plasticizer were added. This will help to improve the cohesiveness of the mix significantly.

19.4.2 Development of Compressive Strength with the Age

Figure 1 displays the development of compressive strength with the age for reinforced EPS concrete. This illustration clearly states that the compressive strength

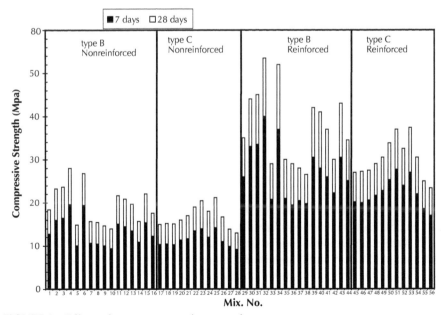

FIGURE 1   Effects of age on compressive strength.

of reinforced EPS concretes in almost all mixes show a continuous increase with age. It is seen that the rate of strength development was greater initially and decreased as the age increased. Comparison of strengths at 7 days revealed that non-reinforced concretes developed almost 40–50% of its 28 day strength. While it is seen that reinforced concretes developed almost 20–40% of the corresponding 28 day strength.

19.4.3   Effect of Density and EPS Volume

Figure 2 and Figure 3 illustrate the compressive strengths of EPS concretes with different plastic densities for reinforced and non-reinforced concrete and the volume content of EPS.

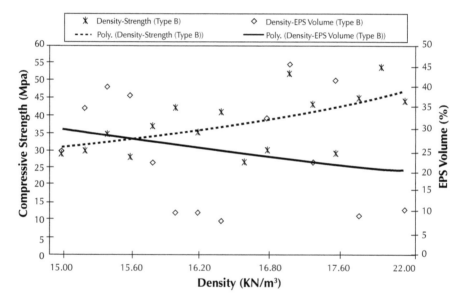

FIGURE 2   Variation of composite density with EPS volume and compressive strength (Type B).

It is appeared that the strength of non-reinforced and reinforced EPS concrete to increase linearly with a decrease in the EPS volume. Furthermore, it is noted that the density of EPS concretes decreased significantly with an increase in the EPS volume. It is clearly observed that the polyamide EPS concrete have high strength in compression, comparing to that non-reinforced concretes.

In some earlier studies reported on EPS concrete, the densities varied from 15 to 20 kN/m^3. At 28 days, it is reported a compressive strength of 10–21MPa for these concretes. In the present study, we obtained the strength of (23.5–37.5) and (26.5–54.0) with use of polyamide 66 yarns.

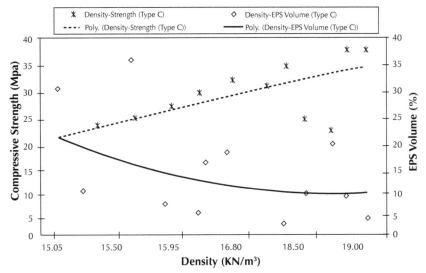

FIGURE 3   Variation of composite density with EPS volume and compressive strength (Type C).

19.4.4 Effect of EPS Bead Size

The plot represented in Figure 4 show an inverse relationship between the concrete strength and the bead size. That means the strength in EPS concrete increased with a decrease in the EPS bead size for the same mix proportions.

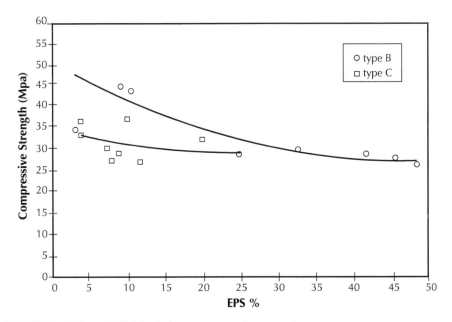

FIGURE 4   Effects of EPS bead size on compressive strength.

19.4.5   Effect of Silica Fume and Polyamide 66 Fibers

From Figure 5 it can be clearly appreciated that in the mixes content fine silica fume the compressive strength can increase significantly (i.e.: mixes no. 4, 6,...). This represents the effects of fine silica fume and polyamide 66 on the compressive strength of EPS concrete. This is clearly visible that fine silica fume can improve the dispersion of EPS in the cement paste and interfacial bonding between EPS and cement paste. Moreover, the compressive strength of polyamide EPS concrete is almost 1.5–2 times non-reinforced EPS concretes.

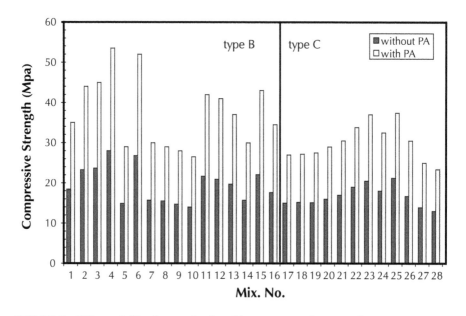

FIGURE 5   Effects of silica fume and polyamide on compressive strength.

19.4.6   Failure Mode

The failure modes of different specimens are presented in Figure 6. Figure 6(a) shows the EPS concrete specimen before loading. The failure mode observed in Figure 6(b) is rather gradual (more compressible), and the specimen is capable of retaining the load after failure, without full disintegration. Figure 6(c) illustrates failure mode in polymer concretes containing EPS with polyamide yarns. This is attributed to the fact that the failure mode of this specimen is near to a typical brittle failure.

19.4.7   Effect of EPS Volume

Figure 7 illustrates the effects of volume content of EPS on the tensile strength of polyamide EPS concretes. This attribute to the fact that there is an inverse relationship between the concrete tensile strength and the EPS volume. This means that the tensile strength of EPS concrete appeared to increase with a decrease in the EPS volume.

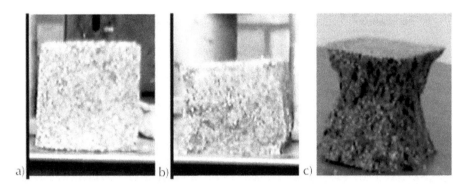

FIGURE 6   The EPS concrete samples with and without polyamide 66 (a) Compressive samples before failure, (b) Observed failure mode in polymer concretes containing EPS without polyamide yarns, and (c) Observed failure mode in polymer concretes containing EPS with polyamide yarns.

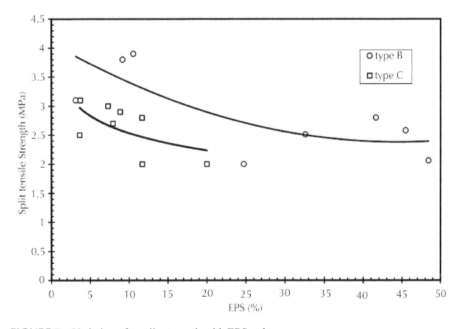

FIGURE 7   Variation of tensile strength with EPS volume.

19.4.8   Effect of EPS Bead Size

From Figure 8, the tensile strength of polyamide EPS concrete (Type B) is 4 MPa whereas, the maximum tensile strength for polyamide EPS concrete (Type C) is only 3.1 MPa. But in general, we can expect an increase of 70% in the tensile strength for the lower bead sizes.

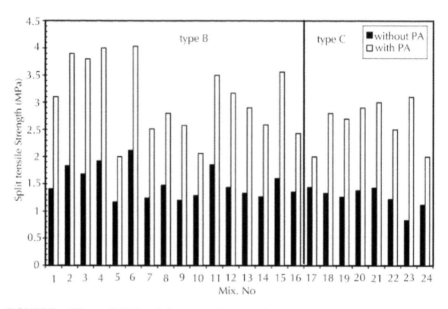

FIGURE 8   Effects of EPS bead size on tensile strength.

19.4.9 Effect of Polyamide 66 Fibers

The effect of polyamide 66 yarns on the tensile strength of EPS concrete is presented in Figure 8. The tensile strength of Polyamide-EPS concrete demonstrated a higher value with respect to EPS concrete. This incremental ratio of strengths is about 60–100%. Furthermore it is observed that the Polyamide-EPS concrete containing smaller beads have higher tensile strength.

19.4.10   Variation of Tensile Strength versus Compressive Strength

Efforts were made to study the variation of tensile strength with the compressive strength (Figure 9). It appears that the tensile strength increased with an increase in compressive strength. The rate of strength increasing in polyamide-EPS concrete containing small beads (Type B) is higher than other.

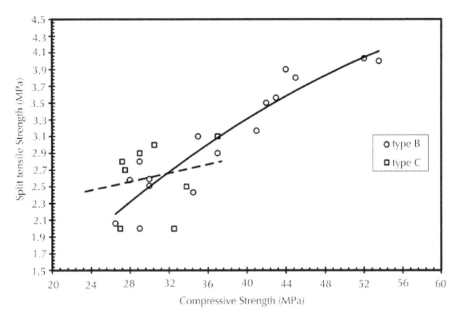

FIGURE 9   Variation of tensile strength with compressive strength.

19.4.11   Failure Mode

The splitting failure mode of the concrete specimens containing non-reinforced EPS aggregates also did not exhibit the typical brittle failure principally expected in conventional concrete as in compressive strength. Consequently, the tensile failure mode observed in the polymer concrete containing polyamide 66 yarns was rather a typical brittle failure which is normally resembles to conventional concrete (Figure 10).

FIGURE 10   The failure mode in split tensile test.

19.4.12   Shrinkage

The most significant factor affecting the shrinkage of concrete is the degree of restraint by the aggregate (i.e. its elastic properties and the volumetric proportion of the paste in the mix). It should be noted that the EPS beads principally offer little hindrance to the shrinkage of the paste. Therefore, it is expected that as the volumetric proportion of the EPS is increased, the shrinkage would increase as well. This is clearly represented in Figure 11. In this study, the drying shrinkage of normal concrete at 90 days was 610 microstrains. Inversely, when the volume content of EPS in the concrete specimen was 20%, the drying shrinkage at 90 days raised to 1,000 microstrain, which represents a disadvantage for the application of EPS.

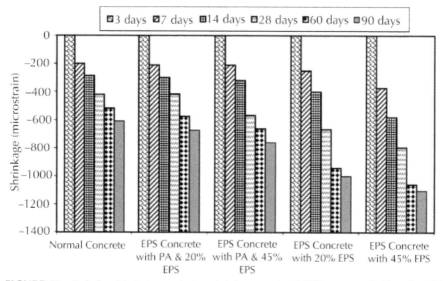

FIGURE 11   Relationship between drying shrinkage strain of EPS concrete (with polyamide and without polyamide) and reference concrete with age.

In a view, it can be seen that polyamide yarns reduced the drying shrinkage of EPS concrete significantly. Even for the case of EPS concrete which contained 20% of its volume by EPS, the drying shrinkage at 90 days was 670 microstrains, which seems to be rather close to normal concrete range.

19.5   CONCLUSION

The use of recycled materials in concrete has been on the increase throughout the world due to conserving resources. The results of this study have shown that use of this locally available low-cost PA waste may provide an economical and effective alternative to concrete aggregates. The chief innovation of this work is the significant ability of polyamide 66 fibers to improve the behavior of lightweight EPS concrete. It is concluded that:

(1) The interaction of Polyamide 66 fibers with EPS concrete did significantly influence the strength. For a 28 days lightweight Polyamide-EPS concrete, we demonstrated strength of around 50MPa while the densities varied from 15 to 20 kN/m³.

(2) The tensile strength of Polyamide-EPS concrete demonstrated a value of about 4 MPa. This incremental ratio of strengths represents an improvement of about 60–100%.

(3) Final absorption of Polyamide-EPS concrete demonstrated a value of 4.5%. This value for EPS concrete was about 8.2%.

(4) The drying shrinkage can be considered as the weakness of EPS concretes. This can be resolved by addition of Polyamide 66 fibers in EPS concrete.
In this study, Polyamide 66 fibers improved the drying shrinkage of EPS concrete significantly. Addition of Polyamide 66 fibers to those EPS concretes containing 20% of EPS demonstrated a drying shrinkage equal to normal concrete. But the presence of polyamide 66 yarns in EPS concretes containing 45% of EPS reduced the drying shrinkage by 30%.

(5) The failure mode of Polyamide-EPS concrete demonstrated a plastic behavior. While this characteristics was brittle in EPS concrete.

KEYWORDS

- **Construction technology**
- **Expanded polystyrene**
- **Municipal solid waste**
- **Polyamide EPS concrete**
- **Waste management**

REFERENCES

1. Muthukumar, M. and Mohsen, D. Studies on polymer concrete based on optimized aggregate mix proportion. *European Polymer Journal*, **40**, 2167–2177 (2004).
2. Gorninski, J. P., Dal Mobin, D. C., and Kazmierczak, C. S. Study of the modulus of elasticity of polymer concrete compounds and comparative assessment of polymer concrete and Portland cement concrete. *Cement and Concrete Research*, **34**, 2091–2095 (2004).
3. ACI Committee 213 R-87. *Guide for Structural Lightweight* Aggregate Concrete, ACI Manual of Concrete Practice, Part 1, American Concrete Institute, Farmington Hills (1987).
4. Short, A. and Kinniburgh, W. *Lightweight Concrete*. (3rd ed.), Applied Science Publishers, London (1978).
5. Sussman, V. *Lightweight plastic aggregate concrete. ACI J*, 321–323 (July, 1975),
6. Bagon, C. and Frondistou-Yannas, S. Marine floating concrete made with polystyrene expanded beads. Mag. *Concr. Res.*, **28**, 225–229 (1976).
7. Babu, K. G. and Babu, D. S. Behavior of lightweight expanded polystyrene concrete containing silica fume. *Cement and Concrete Research*, **33**(5), 755–762 (May, 2003).
8. Al-Manaseer, A. A. and Dalal, T. R. Concrete containing plastic aggregates. *Concr. Int.*, 47–52 (August, 1997)
9. CEB-FIP. Diagnosis and assessment of concrete structures—State of the art report, CEB Bulletin (1989).

10. Roy, R. L., Parant, E., and Boulay, C. Taking into account the inclusions' size in lightweight concrete compressive strength prediction. *Cement and Concrete Research,* (2004).
11. Chen, B. and Liu, J. Properties of lightweight expanded polystyrene concrete reinforced with steel fiber. *Cement and Concrete Research*, **34**, 1259–1263 (2004).
12. Bischoff, P. H. Polystyrene aggregate concrete subjected to hard impact. *Proc. Inst. Civ. Eng., Part* **2**(6), 225– 239 (1990).

20 Application of Polymers in Construction Technology (PART V)

Effects of recycled Plastics on Portland Cement Concrete

A. K. Haghi and G. E. Zaikov

CONTENTS

20.1 INTRODUCTION

For the last decades concrete producers have made wide use of waste or by-product materials in concrete [1]. Proper replacement of these materials in concrete would have two major significant: improving fresh and hardened properties of concrete and minimizing the environmental pollutions due to solid waste disposal. Rubber is one of the materials which are not biodegradable. For solving the disposal of large amount of waste rubber, reuse in concrete industry may be the most feasible application [2].

Recycled waste plastic was also investigated as an additive to Portland cement concrete and as a replacement of aggregate. [6-9] Results showed that the reduction in strength was too great, thus they recommended not replacing more than 20% by volume of the aggregate with recycled plastics.

Also researchers studied on high-strength concrete (HSC) with silica fume that modified with different amounts of crumbed truck tires [10].

20.2 EXPERIMENTAL

Concrete strength is greatly affected by the properties of its constituents and the mixture design parameters. In performance of the experiments, the raw materials used included Portland cement Type 1, mixture of aggregates (coarse and medium), and sand, water and tire fibers. Also silica fume and rice husk ash as high reactive pozzolans were used in this study. Recycled plastics where used as waste materials. Recycled plastics were cut by hand and a cuter in laboratory in form of fiber and chips. They were cut into strips of 30 x 5 x 5 mm, 60 x 5 x 5 mm as fibers, and 10 x 10 x 5 mm and 20 x 20 x 5 mm as chips. The chemical composition of ordinary cement, silica fumes, and rice husk ash are reported in Table 1.

TABLE 1   Chemical composition of materials.

Materials	Ordinary cement	Silica fume	Rice huskash
SiO_2	21.24	91.1	92.1
Al_2O_3	5.97	1.55	0.41
Fe_2O_3	3.34	2	0.21
CaO	62.72	2.24	0.41
MgO	2.36	0.6	0.45
Na_2O	0.13	-	0.08
K2O	0.81	-	2.31
CL	-	-	-
SO_3	1.97	0.45	-
L.O.I	1.46	2.1	-

The concrete mix was rated at 40 Mpa compression strength. A control mix was designed using ACI Standard 211.1 mix design methods. The modified batches designed to be compared with. In the modified batches, 15% by volume of the coarse aggregates was replaced by tires. The mix ratio by weight for control concrete was cement: water: gravel: sand: = 1:0.50:3.50:1.88. The mix ratio by weight for rubberized concrete was cement: water: gravel: waste tire: sand: = 1:0.50:3.40:0.10:1.88 [13].

Also all batches treated with pozzolanic materials as the additional part of cement. In this study 10 and 20% of cement content by weight are selected for addition of pozzolan to the rubberized concrete mixtures [14].

After the concrete was mixed, it was placed in a container to set for 24 hr after that, the specimens were demolded and cured in water at 20°C.

Twenty one batches of 15 cm radius by 30 cm height cylinders were prepared according to ACI specifications. One batch was made without waste tires and pozzolanic addition to be the control while four batches were prepared using waste tire fibers, without any pozzolanic addition.

Remain specimens prepared with waste plastics and additional cementitious part.

Waste plastics measurement of main batches is reported in Table 2. According to this table rubberized modified batches include silica fume and rice husk ash prepared as followed:

TABLE 2   The size of tire in each batch.

Batch number	Waste tire shape	Length (mm)	width (mm)	height (mm)
1	-	-	-	-
2	Tires Fiber	30	5	5
3		60	5	5
4	Tire Chips	10	10	5
5		20	20	5

- ASTM C 39 Standard was used in conducting compressive tests and ASTM C496-86 Standard was used for the split tensile strength tests. Three specimens of each mixture were tested to determine the average strength.
- Flexural toughness determine according to ASTM C1018.
- Compressive and tensile strength of concrete.

The results of the tests for strength performed on the samples in the experiments are shown in Figure 1.

Also its obvious that concrete mixtures with plastic aggregates exhibited higher compressive and lower splitting tensile strengths than other modified batches, but never has passed the regular Portland cement concrete specimens.

Generally there was approximately 40% reduction in compressive strength and 30% reduction in splitting tensile strength when 15% by volume of coarse aggregates were replaced with plastic fibers and chips.

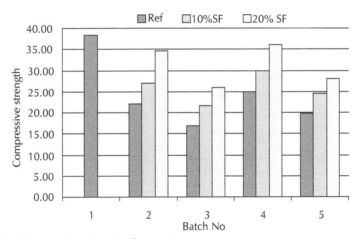

FIGURE 1   Compressive strength of concrete.

20.3   CONCLUDING REMARKS

One of the most crucial environmental issues all around the world is the disposal of the waste materials. According to the large quantity generation of waste tire, it has been a major concern because the waste rubber is not easily biodegradable even after a long-period landfill treatment. The proper disposal of the tires creates an increasing problem that needs to be addressed. Many researchers have investigated the use of recycled plastics products in Civil Engineering constructions. The use of recycled plastics in Portland cement concrete has been the subject of many research projects over the last years. In this way, many experiments were conducted to determine how the properties of concrete were affected by the inclusion of waste tires. Waste tires were used in the form of fibers, chips, ground, and scrap. Studies show that workable plastic waste in concrete mixtures can be made with scrap-plastics.

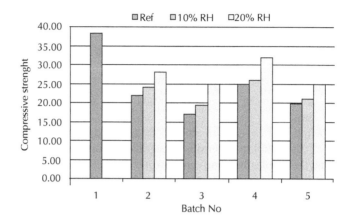

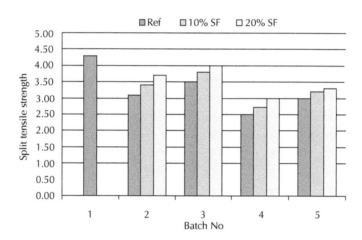

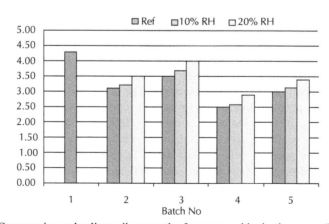

FIGURE 2 Compressive and split tensile strength of concrete with plastic wastes (MPa).

In this study the recycled plastics were further divided into batches with different measurements to determine the effect of size on the properties of new concrete. Totally15% by volume of coarse aggregate was replaced with plastic particles.

Also the concretes were produced by partial addition of cement with silica fume and rice husk ash as high reactive pozzolans. The amount of this addition was varying at 10 and 20% by weight of cement to find the effects of pozzolans to improve the mechanical properties of Rubcrete. So, 21 concrete mixtures were cast and tested for compressive and splitting tensile strength, static modulus of elasticity and slump in accordance to ASTM standards.

As results there was a noticeable decline in the compressive and tensile strength of new concrete. However, the addition of silica fume and rice husk ash into the matrix improved the mechanical properties of the new concretes and diminished the rate of strength loss. Moreover there was an increase in the toughness. It was concluded that recycled plastics were suitable as additives produced the highest toughness.

KEYWORDS

- **Concrete strength**
- **Plastic fibers and chips**
- **Portland cement concrete**
- **Pozzolans**
- **Waste materials**

REFERENCES

1. Refined, J. Development of non-traditional glass markets. *Resour. Recycl.*, 18–21 (1991).
2. Uchiyama, J. Long-term utilization of the glass reasphalt pavement. *Pavement*, 3–89 (1998).
3. Johnson, C. D. Waste glass as coarse aggregate for concrete. *J. Test. Eval.*, **2**(5) 344–350 (1974).
4. Masaki, O. *Study on the hydration hardening character of glass powder and basic physical properties of waste glass as a construction material.* Asahi Ceramic Foundation Annual Research Report, 1995.
5. Park, S. B. *Development of Recycling and Treatment Technologies for Construction Wastes.* Ministry of Construction and Transportation, Seoul (2000).
6. Swamy, R. N. *The Alkali–Silica Reaction in Concrete.* Van Nostrand Reinhold, New York, (1992).
7. Naohiro, S. *The strength characteristics of mortar containing glass powder.* The 49th Cement Technology Symposium. JCA, Tokyo, (1995) 114–119.
8. Kyoichi, I. and Atobumi, K. Effects of glass powder on compressive strength of cement mortar, College of Engineering, Kantou Gakuin University. *Study Report*, **40**(1) 13–17 (1996).
9. Bentur, A. and Mindess, S. *Fiber reinforced cementitious on durability of concrete.* Barking, Elsevier (1990).
10. Balaguru, P. N. and Shah, S. P. *Fiber reinforced cement composites.* McGraw-Hill, Inc, New York, p. 367 (1992).
11. Park, S. B. *Development of Energy Conserving High Performance Fiber Reinforced Concrete and Instruction for Design and Construction for the Reinforced Concrete*, Ministry of Construction and Transportation, Seoul (1998).

21 Application of Polymers in Construction Technology (PART VI)

Effects of Recycled Textile fibers on reinforcement of cement

A. K. Haghi and G. E. Zaikov

CONTENTS

21.1 INTRODUCTION

Industrial waste of different origins and nature can be used as unconventional constituents for the preparation of new blended cements. This study collects different researches previously carried out with the aim to highlight the feasibility of this recycling route that can be considered highly rewarding for the cement industry. Chemical, physical, and mechanical properties of the new blended cements are reviewed and compared with the requirements set by EN 197-1 for common cements.

In this work industrial waste and relevant new blended cement constituted by 25 wt% waste and 75 wt% CEM I, previously separately investigated are collectively reported and compared.

A chemical and mineralogical characterization of the materials was then carried out to establish the main oxide constituents and mineralogical phases. The results are summarized in Tables 2 and 3: SiO_2 is the main constituent of MW, GR, and PR, but while MW has the typical chemical-mineralogical composition of soda-lime glass (about 10 and 13 wt% content of CaO and Na_2O, respectively, and amorphous silica phase), GR and PR are also rich in Al_2O_3 (15–20 wt%) and ZrO_2 (1–3 wt%), minerals such as zircon, quartz and albite deriving from glazing and porcelain stoneware body,

TABLE 1 Cement classification according to EN 197-1 (wt%).

Main types	Notations		Main Constituents								
			Clinker K	Blast Furnace slag S	Silica fume D	Pozzolana		Fly ash		Burnt shale T	Limestone
						nat P	natural calcined Q	Sil. V	Calc. W		L LL
CEM I	Portland Cement	CEM I	95–100	–	–	–	–	–	–	–	– –
	Portland slag cement	CEM II/ A-S	80–94	6–20	–	–	–	–	–	–	– –
		CEM II/ B-S	80–94	21–35	–	–	–	–	–	–	– –
	Portland silica-fume cement	CEM II/A-D	90–94	–	6–10	–	–	–	–	–	– –
		CEM II/A-P	80–94	–	–	6–20	–	–	–	–	– –
		CEM II/B-P	65–79	–	–	21–35	–	–	–	–	– –
	Portland pozzolan cement	CEM II/A-Q	80–94	–	–	–	6–20	–	–	–	– –
		CEM II/B-Q	65–79	–	–	–	21–35	–	–	–	– –
		CEM II/A-V	80–94	–	–	–	–	6–20	–	–	– –
	Portland fly-ash cement	CEM II/B-V	65–79	–	–	–	–	21–35	–	–	– –
CEM II		CEMII/A-W	80–94	–	–	–	–	–	6–20	–	– –
		CEMII/ B-W	65–79	–	–	–	–	–	21–35	–	– –
	Portland burnt shale cement	CEMII/ A-T	80–94	-	-	-	-	-	-	6-20	- -
		CEMII/ B-T	65–79	-	-	-	-	-	-	21-35	- -
	Portland limestone cement	CEM II/A-L	80–94	-	-	-	-	-	–	6–20	–
		CEM II/B-L	65–79	–	–	–	–	–	–	21–35	–
		CEM II/A-LL	80–94	–	–	–	–	–	–	–	– 6–20
		CEM II/B-LL	65–79	–	–	–	–	–	–	–	– 21–35
	Portland composite cement	CEM II/A-M	80–94			6–20					
		CEM II/B-M	65–79			21–35					

TABLE 1 *(Continued)*

Main types	Notations		Clinker K	Blast Furnace slag S	Silica fume D	Pozzolana				Fly ash		Burnt shale T	Limestone	
						nat P	natural calcined Q	Sil. V	Calc. W				L	LL
CEM III	Blast furnace cement	CEMIII/A	35–64	36–65	–	–	–	–	–	–	–	–	–	
		CEMIII/B	20–34	66–80	–	–	–	–	–	–	–	–	–	
		CEMIII/C	5–19	81–95	–	–	–	–	–	–	–	–	–	
CEM IV	Pozzolan cement	CEMIV/A	65–89	–		11–35				–	–	–		
		CEMIV/B	45–64			36–55				–	–	–		
CEM V	Composite cement	CEMV/A	40–64	18–30	–	18–30				–	–	–	–	
		CEMV/B	20–38	31–50	–	31–50				–	–	–	–	

respectively. The main constituents of LS are CaO (55 wt%), SiO_2 (24 wt%), and Al_2O_3 (13wt%) with mineralogical crystalline phases as calcium silicates and aluminates.

TABLE 2 Chemical analysis (main oxide, wt%) of the investigated industrial waste (x-ray fluorescence spectrometer, XRF, PW 1414, Philips).

Oxide	MW (%)	LS (%)	GR (%)	PR (%)
SiO_2	71.04	23.85	52.36	62.19
Al_2O_3	2.02	13.69	19.37	15.75
TiO_2	–	0.20	0.45	0.34
Fe_2O_3	0.35	3.85	0.84	0.59
CaO	10.58	55.24	5.73	2.24
MgO	1.75	3.22	2.43	6.75
K_2O	0.75	0.22	1.32	1.46
Na_2O	13.52	0.31	3.90	3.71
ZrO_2	n.d.	<0.10	3.01	1.19
MnO	n.d.	0.32	<0.10	<0.10
ZnO	n.d.	<0.10	0.99	0.12
BaO	n.d.	<0.10	0.54	<0.10

TABLE 3   Main mineralogical phases of the investigated industrial waste (x-ray diffractometer with Ni-filtered Cu Kα ($\lambda = 1.54$ Å) radiation, XRD PW 1,840, Philips).

	MW	LS	GR	PR
Crystalline phase	n.d.	olivine (γ–C_2S), ghel-enite, mayenite, iron silicates, iron magnesium calcium silicate	quartz, zircon, albite	quartz, albite calcian
Amorphous phase	silica	n.d.	silica	silica

Table 4 reports the values determined according to EN 196-2 [16] for the investigated materials and OPC CEM I 52.5R. MW, LS, and GR values agree with all the limits sets, whereas PR exhibits a higher Cl⁻ content and slightly overcomes the limit LOI value. Cl⁻ derives from the salts ($AlCl_3$, $FeCl_3$) used as flocculants in the separation process and from the magnesium chloride matrix of abrasive tools used for polishing. The LOI value is mainly related to the presence of calcite phase and SiC, coming from the abrasive tools. However, both the data can be acceptable as PR is going to be used mixed with CEM I 52.5 R (25 and 75% respectively), hence the overall Cl⁻ % and LOI will be inside the limits required for cement.

TABLE 4   Chemical analysis results of the investigated industrial waste and CEM I 52.5 R (average of 2 measurements). Limits set by EN 197-1 for cement are also reported.

	MW	LS	GR	PR	CEM I 52.5 R	Limits set by EN 197-1 for cement
Chloride (wt %) EN 196-2	0.04	≤ 0.01	≤ 0.01	0.24	0.04	≤ 0.1
Sulfate (SO_3 wt %) EN 196-2	≤ 0.01	0.38	0.06	0.09	0.65	≤ 3.5
Loss of Ignition (wt %) EN 196-2	0.8	2.7	2.9	5.2	3.6	≤ 5.0

Table 5 reports the results obtained on the new blended cement based on the investigated waste, having the general composition: 75 wt% CEM I 52.5 R + 25 wt% waste.

TABLE 5   Physical properties of the investigated binder based on industrial waste and CEM I 52.5 R (average of 2 measurements). Limits set by EN 197-1 for cement are also reported.

Binder		Initial setting time	Soundness
		(min)	(mm)
CEM I 52.5 R		107	0.2
75% CEM I 52.5 R + 25% MW		134	0.3
75% CEM I 52.5 R + 25% LS		106	0.4
75% CEM I 52.5 R + 25% GR		117	0.2
75% CEM I 52.5 R + 25% PR		105	0.2
Limits set by EN 197/1 for cement	32.5 R	≥ 75	≤ 10
	42.5 R	≥ 60	
	52.5 R	≥ 45	

Recycled Glass

Glass has been used as aggregate in road constructions, building, and masonry materials [13]. Recent studies have shown that reuse of very finely ground waste glass in concrete has economical and technical advantages [11-14].

Recycled Fibers

A great amount of fibrous textile waste is discarded into landfills each year all over the world. More than half of this waste is from carpets, which decays at a very slow rate and which is difficult to handle in landfills.

21.2   EXPERIMENTAL

Materials used included ordinary Portland cement type 1, standard sand, silica fume, glass with tow particle size, rice husk ash (RH), tap water and finally fibrillated polypropylene fibers.

The fibers included in this study were monofilament fibers obtained from industrial recycled raw materials that were cut in factory to 6 mm length. Properties of waste Polypropylene fibers are reported in Table 6.

TABLE 6   Properties of polypropylene fibers.

Property	Polypropylene
Unit weight [g/cm3]	0.9 – 0.91
Reaction with water	Hydrophobic
Tensile strength [ksi]	4.5 – 6.0
Elongation at break [%]	100 – 600
Melting point [°C]	175
Thermal conductivity [W/m/K]	0.12

Also the silica fume (SF) and RH contain 91.1% and 92.1% SiO_2 with average size of 7.38 μm and 15.83 μm respectively were used. The chemical compositions of all pozzolanic materials containing the reused glass, SF and RH were analyzed using an x-ray microprobe analyzer and listed in Table 7.

In accordance to ASTM C618, the glass satisfies the basic chemical requirements for a pozzolanic material especially clean glass. To satisfy the physical requirements for fineness, the glass has to be grounded to pass a 45 μm sieve.

TABLE 7   Chemical composition of materials.

	Content (%)		
Oxide	Glass C	Silica fume	Rice husk ash
SiO_2	72.5	91.1	92.1
Al_2O_3	1.06	1.55	0.41
Fe_2O_3	0.36	2	0.21
CaO	8	2.24	0.41
MgO	4.18	0.6	0.45
Na_2O	13.1	–	0.08
K_2O	0.26	–	2.31
Cl	0.05	–	–
SO_3	0.18	0.45	–
L.O.I	–	2.1	–

To obtain this aim recycled windows clean glass was crushed and grinded in laboratory, and sieved the ground glass to the desired particle size. To study the particle size effect, two different ground glasses were used, namely:

- *Type I*: ground glass having particles passing a #80 sieve (180 μm);
- *Type II*: ground glass having particles passing a #200 sieve (75 μm).

In addition the particle size distribution for two types of ground glass, silica fume, RH and ordinary Portland cement were analyzed by laser particle size set, have shown in Figure 1. As it can be seen in SF has the finest particle size. According to ASTM C618, fine ground glasses under 45 μm qualify as a pozzolan due to the fine particle size. Moreover glass type I and II respectively have 42% and 70% fine particles smaller than 45μm that causes pozzolanic behavior.

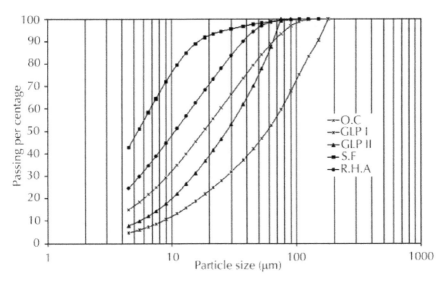

FIGURE 1   Particle size distribution of ground waste glass Type I, II, silica fume, RH and ordinary cement.

For the present study, twenty batches were prepared. Control mixes was designed containing standard sand at a ratio of 2.25:1 to the cement in matrix. A partial replacements of cement with pozzolans include ground waste glass (GI, GII), SF and RH were used to examine the effects of pozzolanic materials on mechanical properties of PP reinforced mortars at high temperatures. The amount of pozzolans which replaced was 10% by weight of cement which is range that is most often used.

Meanwhile, polypropylene fibers were used as addition by volume fraction of specimens. The reinforced mixtures contained PP fiber with three designated fiber contents of 0.5, 1, and 1.5% by total volume.

In the plain batches without any fibers, water to cementitious ratio of 0.47 was used whereas in modified mixes (with different amount of PP fibers) it changed to 0.6 due to water absorption of fibers. The mix proportions of mortars are given in Table 8.

The strength criteria of mortar specimens and impacts of polypropylene fibers on characteristics of them were evaluated at the age of 60 days.

In our laboratory, the test program mix conducted as follows:

1.   The fibers were placed in the mixer.
2.   Three-quarters of the water was added to the fibers while the mixer was running at 60 rpm; mixing continues for one minute.
3.   The cement was gradually the cement to mix with the water.
4.   The sand and remaining water were added, and the mixer was allowed to run for another two minutes.

TABLE 8 Mixter properties.

Batch No	Sand/c	w/c	Content (by weight)					PP fibers (by volume)	batch No	sand/c	w/c	Content (by weight)					PP fibers (by volume)
			O.C	GI	GII	SF	RH					O.C	GI	GII	SF	RH	
1	2.25	0.47	100	–	–	–	–	0	11	2.25	0.6	100	–	–	–	–	1
2	2.25	0.47	90	10	–	–	–	0	12	2.25	0.6	90	10	–	–	–	1
3	2.25	0.47	90	–	10	–	–	0	13	2.25	0.6	90	–	10	–	–	1
4	2.25	0.47	90	–	–	10	–	0	14	2.25	0.6	90	–	–	10	–	1
5	2.25	0.47	90	–	–	–	10	0	15	2.25	0.6	90	–	–	–	10	1
6	2.25	0.6	100	–	–	–	–	0.5	16	2.25	0.6	100	–	–	–	–	1.5
7	2.25	0.6	90	10	–	–	–	0.5	17	2.25	0.6	90	10	–	–	–	1.5
8	2.25	0.6	90	–	10	–	–	0.5	18	2.25	0.6	90	–	10	–	–	1.5
9	2.25	0.6	90	–	–	10	–	0.5	19	2.25	0.6	90	–	–	10	–	1.5
10	2.25	0.6	90	–	–	–	10	0.5	20	2.25	0.6	90	–	–	–	10	1.5

After mixing, the samples were casted into the forms 50×50×50 mm for compressive strength and 50×50×50 mm for flexural strength tests. All the moulds were coated with mineral oil to facilitate demoulding. The samples were placed in two layers. Each layer was tamped 25 times using a hard rubber mallet. The sample surfaces were finished using a metal spatula. After 24 hr, the specimens were demoulded and cured in water at 20°C.

The heating equipment was an electrically heated set. The specimens were positioned in heater and heated to desire temperature of 300 and 600°C at a rate of 10–12°C/min. After 3 hr, heater turned off. It was allowed to cool down before the specimens were removed to prevent thermal shock to the specimens. The rate of cooling was not controlled. The testes to determine the strength were made for all specimens at the age of 60 days. At least three specimens were tested for each variable.

21.3 DISCUSSIONS AND RESULTS

21.3.1 Density
The initial density of specimens containing polypropylene fibers was less than that of mixes without any fibers. Density of control mixes without any replacement of cement at 23,300 and 600°C are reported in Table 9. According to the results, density decrease of fiber reinforced specimens was close to that of plain ones. The weight of the melted fibers was negligible. The weight change of mortar was mainly due to the dehydration of cement paste.

TABLE 9   Density of control specimens.

Heated at	23°	300°	600°	PP fibers
Density (gr/cm³)	2.57	2.45	2.45	0%
Density (gr/cm³)	2.50	2.36	2.36	0.5%
Density (gr/cm³)	2.44	2.28	2.27	1%
Density (gr/cm³)	2.41	2.23	2.21	1.5%

21.3.2 Flexural Strength
The specimens were used for flexural testes were 50×50×200 mm. The results of plain specimens and samples containing 1.5% fibers are shown in Table 10.The heat resistance of the flexural strength appeared to decrease when polypropylene fibers were incorporated into mortar. This is probably due to the additional porosity and small channels created in the matrix of mortar by the fibers melting like compressive strength. However the effect of the pozzolans on flexural strength is not clear but it seems that SF and glass types II have better impact on strength in compare with control specimens than RH and glass type I.

TABLE 10   Flexural strength of control samples with 0 and 1.5% fibers.

Batch	Flexural Strength (Mpa)			PP fibers
No	23°	300°	600°	(by volume)
1	4	3.1	2.8	0
2	3	2.4	2.1	0
3	3.7	3.2	2.7	0
4	3.9	3.4	2.7	0
5	3.3	2.8	2.6	0
16	2.2	1	0.7	1.5
17	1.6	0.8	0.6	1.5
18	2	1.1	0.8	1.5
19	2.4	1.3	0.8	1.5
20	1.8	1	0.6	1.5

21.4   CONCLUSION

In this chapter it is shown that the industrial waste of different origins and nature can be used as unconventional constituents for the preparation of new blended cements. This study collects different researches previously carried out with the aim to highlight the feasibility of this recycling route that can be considered highly rewarding for the cement industry. Chemical, physical and mechanical properties of the new blended cements are reviewed and compared with the requirements set by EN 197-1 for common cements.

KEYWORDS

- **Cement industry**
- **Industrial waste**
- **Physical properties**
- **Polypropylene fibres**
- **Pozzolanic material**

REFERENCES

1. Wei, M. S. and Huang, K. H. Recycling and reuse of industrial waste in Taiwan. *Waste Management,* **21,** 93–97 (2001).
2. Mildness, S., Young, J. F., and Darwin, D. *Concrete.* Prentice Hall, New Jersey (2003).
3. ACI Committee 232, Use of fly ash in concrete (ACI 232.2R-96), *ACI Manual of Concrete Practice, Part 1,* American Concrete Institute, Farmington Hills, (2001).
4. ACI Committee 232, Use of Raw or Processed Natural Pozzolans in Concrete (ACI 232.1R-00), American Concrete Institute, Farmington Hills, (2000).

5. ACI Committee 233, Ground, Granulated Blast Furnace Slag as a Cementitious Constituent in Concrete (ACI 233R-95). *ACI Manual of Concrete Practice, Part 1.* American Concrete Institute, Farmington Hills, (2001).

6. ACI Committee 234, Guide for Use of Silica Fume in Concrete (ACI 234R-96). *ACI Manual of Concrete Practice, Part 1.* American concrete Institute, Farmington Hills, (2001).

7. Harrison, W. H. Synthetic aggregate sources and resources, *Concrete,* **8**(11), 41–44 (1974).

8. Johnston, C. D. Waste glass as coarse aggregate for concrete. *Journal of Testing and Evaluation,* **2**(5), 344–350 (1974).

9. Meyer, C., Baxter, S., and Jin, W. Potential of waste glass for concrete masonry blocks. In *Materials for a New Millennium.* K. P. Chong (Ed.). Proceedings of ASCE Materials Engineering Conference, Washington, DC, pp. 666–673 (1996).

10. Polley, C., Cramer, S. M., and Cruz, R. V. Potential for using waste glass in Portland cement concrete. *Journal Materials in Civil Engineering,* **10**(4), 210–219 (1998).

11. Shao, Y., Lefort, T., Moras, S., and Rodriguez, D. Studies on concrete containing ground waste glass. *Cement and Concrete Research,* **30**(1), 91–100 (2000).

12. Shayan, A. and Xu, A. Value-added utilisation of waste glass in concrete. *Cement and Concrete Research,* **34**, 81–89 (2004).

13. Reindl, J. Report by recycling manager. *Dane County.* Department of Public Works, Madison (1998).

14. Park, S. B. and Lee, B. C. Studies on expansion properties in mortar containing waste glass and fibers. *Cement and Concrete Research,* **34**, 1145–1152 (2004).

15. Panchakarla, V. S. and Hall, M. W. Glascrete—disposing of non-recyclable glass. In *Materials for a New Millennium,* K. P. Chong (Ed.). Proceedings of ASCE Materials Engineering Conference, Washington, DC, pp. 509–518 (1996).

16. Meyer, C., Baxter, S., and Jin, W. Alkali-silica reaction in concrete with waste glass as aggregate. *Materials for a New Millennium.* In K. P. Chong (Ed.). Proceedings of ASCE Materials Engineering Conference, Washington, DC, pp. 1388–1394 (1996).

17. Panchakarla, V. S. and Hall, M. W. Glascrete—disposing of non-recyclable glass. In *Materials for a New Millennium,* K.P. Chong (Ed.). Proceedings of ASCE Materials Engineering Conference, Washington, DC, pp. 509–518. (1996).

18. Bentur, A. and Mindess, S. Fiber reinforced cementitious on durability of concrete. Barking Elsevier (1990).

19. Wang, Y. and Wu, H. C. Concrete reinforcement with recycled fibers. *Journal of Materials in Civil Engineering* (2000).

20. Flynn L. Auchey and Piyush K. Dutta Use of recycled high density polyethylene fibers as secondary reinforcement in concrete subje cted to severe environment. *Proceedings of the International Offshore and Polar Engineering Conference* (1996).

21. Wu, H. C., Lim, Y. M., and Li, V. C. Application of recycled tyre cord in concrete for shrinkage crack control. *Journal of Materials Science Letters,* **15**(20), 1828–1831 (October 15, 1996).

22. Naaman, A. E., Garcia, S., Korkmaz, M., and Li, V. C. Investigation of the use of carpet waste PP fibers in concrete. *Proceedings of the Materials Engineering Conference* **1**, 799–808 (1996).

23. Balaguru, P. N and Shah, S. P. *Fiber reinforced cement composites.* McGraw Hill Inc, New York, p. 367 (1992).

24. Kumar, S., Polk, M. B., and Wang, Y. Fundamental studies on the utilization of carpet waste, presented at the SMART (Secondary Materials & Recycled Textiles, An International Association) Mid-Year Conference, Atlanta, GA (July, 1994).

25. Terro, M. J. and Hamoush, S. A. Behavior of confined normal-weight concrete under elevated temperature conditions. *ACI Materials Journal,* **94**(2), 83–89 (1997).

26. Terro, M. J. and Sawan, J. Compressive strength of concrete made with silica fume under elevated temperature conditions. *Kuwait Journal of Science and Engineering,* **25**(1), 129–144 (1998).

27. Kalifa, P., Menneteau, F. D., and Quenard, D. Spalling and pore pressure in HPC at high temperatures. *Cement and Concrete Research,* **30**(12), 1915–1927 (2000).

28. Saad, M., Abo-El-Ein, S. A., Hanna, G. B., and Kotkata, M. F. Effect of temperature on physical and mechanical properties of concrete containing silica fume. *Cement and Concrete Research,* **26**(5), 669–675 (1996).

29. Xu, Y, Wong, Y. L., Poon, C. S., and Anson, M. Impact of high temperature on PFA concrete. *Cement and Concrete Research*, **31**(7), 1065–1073 (2001).
30. Kodur, V. K. R., Wang, T. C., and Cheng, F. P. Predicting the fire resistance behavior of high strength concrete columns. *Cement and Concrete Composites*, **26**(2), 141–153 (2004).
31. Poon, C. S., Azhar, S., Anson, M., and Wong, Y. L. Performance of metakaolin concrete at elevated temperatures. *Cement and Concrete Composites*, **25**(1), 83–89 (2003).
32. Savva, A., Manita, P., and Sideris, K. K. Influence of elevated temperatures on the mechanical properties of blended cement concretes prepared with limestone and siliceous aggregates. *Cement and Concrete Composites*, **27**(2), 239–248 (2005).

Index

A

Afonicheva, O. V., 121–137
Aggregation parameter, 17–18
Akhmetshina, L. F., 25–38
Aloev, V. Z., 175–178
Anionic mechanism, 122
Anionic ring-opening polymerization of e-caprolactam (APCL), 121
  influence of
    IL concentration on, 124–125
    IL nature on, 124
Anisotropic MMT particles, 67
Arburg Allounder 305-210-700 molding machine, 40
Archibald method, 167
Aspergillus niger, 141
ASTM C 143-98, 215
ASTM C 305, 215
ASTM C 494, 232
ASTM C 496-89, 215, 232
ASTM C618, 181, 182, 188
ASTM C 642-82, 232
ASTM C989, 188
ATP, high energy bonds transformation, 87
  anion-anionic distances, 98
  ATP synthesis, 99–100
  biological systems, 93, 99
  biomolecules, 97
  bond energy, 93, 98–100
  crystalline ionic, 93
  glycolysis, 100
  harmonic oscillator, 90
  ionized phosphate, 100
  isomorphism, 91–93, 99
  Lagrangian equation, 87
  oscillations, 89–90
  parameter methodology, 92
  parameters equations, 90
  phosphorus anhydrite, 97
  quantum number, 88
  quasi-classical, 89
  sesquialteral bond, 98–99, 106
  spatial-energy, 87–88, 93, 99
  stable biostructure, 99
  Van der Waals forces, 97–100
  wave properties, 89, 91

B

Bacteria, 139, 141
"Binary hooking," 42
Bioactive compounds, 75
Biocide polyorganosiloxanes, 139
Biological damage and protection of materials, 113
  adult insect, 114, 116, 119
  autumn wood, 116
  book beetle, 118
  chemical control, 118
  clock ticking, 114
  constructive, 118
  cross hollow, 114
  curious facts, 114
  destructive, 118
  drugstore beetle, 118
  embryonic development, 115
  feigns dead, 115
  firewood, 118
  hazardous enemies, 114
  hexachlorane, 119
  humidity, 115–117
  imago, 114
  killer agant, 119
  larva, 114–116, 119
  oviposit, 115
  ovum, 114
  paraffin, 119
  phylogenic or zoogenic food, 116
  prophylactic measures, 118
  pupa, 114, 116, 119
  springwood, 116
  symbionts, 116
  temperature, 114–117, 122–123, 130
  turpentine, 119
  wax, 119
  wood fretter larvae, 114, 116, 119

Milton Keynes UK
Ingram Content Group UK Ltd.
UKHW031145141024
449569UK00024B/1048